# *It's All a* GAME

**Also by Tristan Donovan**

*Replay: The History of Video Games*

*Fizz: How Soda Shook Up the World*

*Feral Cities: Adventures with Animals in the Urban Jungle*

# It's All a GAME

**THE HISTORY OF BOARD GAMES**
*from* **MONOPOLY**
*to* **SETTLERS OF CATAN**

Tristan Donovan

Thomas Dunne Books
St. Martin's Press New York

THOMAS DUNNE BOOKS.
An imprint of St. Martin's Press.



www.thomasdunnebooks.com
www.stmartins.com

Book design by Michelle McMillian

Library of Congress Cataloging-in-Publication Data

Names: Donovan, Tristan, author.
Title: It's all a game : the history of board games from Monopoly to settlers of Catan / Tristan Donovan.
Description: New York : Thomas Dunne Books, 2017.
Identifiers: LCCN 2017001975| ISBN 9781250082725 (hardcover) | ISBN 9781250082732 (e-book)
Subjects: LCSH: Board games—History. | Board games—Social aspects. | BISAC: HISTORY / Social History.
Classification: LCC GV1312.D66 2017 | DDC 794—dc23
LC record available at https://lccn.loc.gov/2017001975

Our books may be purchased in bulk for promotional, educational, or business use. Please contact your local bookseller or the Macmillan Corporate and Premium Sales Department at 1-800-221-7945, extension 5442, or by e-mail at MacmillanSpecialMarkets@macmillan.com.

First Edition: May 2017

10 9 8 7 6 5 4 3 2 1

To my sister Jade,

the queen of overturned Monopoly boards

# CONTENTS

# It's All a GAME

# INTRODUCTION

## *The Birth of a New Gaming Era*

"Hey, bud! Wanna game?"

It's a balmy Saturday in Manhattan and New Yorkers are unwinding in Washington Square Park. As sunbathers read books and doze in the sunshine, a jazz ensemble is producing sounds so disorientating that it's hard to tell if the musicians are playing together or trying to drown one another out. Next to them a man is blowing enormous soap bubbles to the delight of children while tourists recoil from the guy encouraging them to pose for a photo with the mean-looking snake draped around his shoulders.

And over in the park's shady southeast corner is the Chess Plaza, and the man who's urging me to join him for a game. He beckons me closer with a wave of his hand, hoping to grab me before any of the other players, who, like him, are sitting alone at the tables trying to catch the interest of passersby. It seems like no one's up for chess today.

"I'm a hustler, I play for money," says the man as I approach, his dirty white baseball cap casting a shadow over his wrinkled face. "How's twenty dollars sound?"

Charles has been playing chess in the park for decades. So long in fact that he can't remember not coming here. "I'm sixty-seven, born in New York City. In the park on and off, I guess, my whole life," he tells me. Charles caught the chess bug at the age of eleven after his father bought him a set. "Then I came to the park one time and I saw them playing and went, 'Wow! That's so much fun.' They were good players . . . dissing each other and the sound of the pieces, the clack, the knock . . . Their conversation was fantastic."

From that moment on Charles has been hooked. "There's so many different variations to chess it seems like there are infinite moves," he explains. "I like chess also because it passes the time real quick. Hours can go by and it seems like nothing."

When Charles started coming here in the early 1960s, Washington Square Park was chess central. Its chess scene cut through racial and class divides. Here it didn't matter if you were white or black, Wall Street millionaire or joint-smoking hippie, tourist or native New Yorker. All that mattered was the game and how you played.

In years gone by, grandmasters like Bobby Fischer and Roman Dzindzichashvili would play before huddled crowds and celebrities ranging from Stanley Kubrick to Chris Rock. At the center of it all were the hustlers, the real stars of the Washington Square Park chess scene, people with nicknames like Russian Paul and Little Daddy. Many were unemployed, getting by with the dollars they made at the chessboards, and the best became local celebrities thanks to their showmanship and entertaining banter.

Charles didn't start out as a hustler. "I just wanted to play chess. Then the park changed, it got very much into the gambling aspect. Everybody was gambling so I had to move with the crowd, but I'm a good hustler," he laughs.

Indeed he is. We're not even three minutes into our first game and he is already tearing me to shreds. "Now you're finished, you made this too easy for me," says Charles as he scoops up one of my bishops. "You're in so much bad shape."

If Charles initially thought he had caught a fish, as Chess Plaza regulars refer to weak players who they can make money from, now he knew he had. A few minutes later, I'm cooked. "Checkmate," he cries.

As we start setting up for another game, the other chess players continue hawking in vain. "Want a game?" one calls out to a passing man. "No? You don't play chess! Would you like to learn? No? Okay, have a nice day."

The Washington Square chess scene isn't what it used to be, says Charles with a hint of sadness. There used to be more players and crowds. Many of the players who used to congregate in Washington Square are now in Union Square Park. They decamped there when the parks department began renovating the Chess Plaza in the early 2010s, playing on makeshift tables built from plastic crates and slabs of cardboard. Many never came back. Union Square proved a busier and more profitable place to play.

There are other signs of decline in what used to be called Manhattan's Chess District. Even the Village Chess Shop, once a must-see destination for chess lovers, has closed. Located a short half block from the park at 230 Thompson Street, the Village Chess Shop opened in 1972. On the sidewalk outside players would battle day and night on the store's ramshackle chairs and tables while people paused to marvel at the clutter of strange and bizarre chess sets on display in the windows. It became a New York icon, attracting stars like Woody Allen, Heath Ledger, David Lee Roth, and John Lennon.

The Village Chess Shop closed in 2012, telling its Facebook followers that "the chess shop became more of a curiosity or portrait than a viable retail environment." Its nemesis the Chess Forum, which opened right across the street in 1995, is still there but it too is struggling as online retailers erode its sales.

The Village Chess Shop eventually resurfaced in a smaller venue just around the corner, but all the same it is tempting to see the steady decline of the Washington Square chess scene as a sign of the times—proof that board games are on their way out, pushed aside in our rush to embrace the glitzy thrills available on our PlayStations and smartphones.

Board games had a good run. People first began playing them centuries, maybe even millennia, before the development of the written word and they have been with us ever since, but "the times they are a-changin'" and all that.

Yet right there, in the same retail space where the Village Chess Shop once resided, a different kind of board game business is thriving. The new resident of 230 Thompson Street is the Uncommons and it is one of a legion of board game cafés that have opened in cities across the world in recent years.

Inside the young crowd, whose ages range from early teens to thirtysomethings, hums with excitement. Today, the Uncommons is hosting a special event: a prerelease game of Magic Origins, the latest deck for the wildly popular trading card game Magic: The Gathering. While the prospect of an early taste of Magic Origins attracted today's crowd, the café's owner Greg May tells me that being this busy is not unusual. "It will be quiet until one p.m. on a Saturday, we'd usually have five or six people, and then at one p.m. it starts filling up fast and by two o'clock there's usually a one-hour wait," he says.

That the Uncommons has taken the place of the Village Chess Shop is no coincidence. May spent months trying to find the landlord so he could claim the space for his new board game café. "I coveted the spot but couldn't get any information on it. Then one day I was walking by and there was a sign up in this space finally listing the Realtor's name and that was basically it," he says. "I love the history. I had been to the Village Chess Shop before it had closed and really appreciate the community and traditions this space holds."

Nonetheless overlap between the Uncommons and Chess Plaza crowd is limited. "We've definitely shifted away from that chess audience," says May. "Our top customer base are the students, the NYU students on campus. They come for board game nights, they come for tournaments, they come on dates. We're a big date spot. Weirdly, to me at least, we've become really popular in the Hasidic Jewish community. They see us as a fun, family-friendly, fairly safe place to take someone out on a very structured date, where it's public and not too dark. We also have a ton of families who come, especially on the weekends. Parents bring in their kids to introduce them to chess or Scrabble."

The Uncommons' business model is straightforward. Players pay five to ten dollars plus tax per person and then pick a game to play from a collection that boasts close to eight hundred options. The collection covers the gamut of board gaming. Among them are ancient perennials like chess, family favorites like Monopoly, emerging classics such as Ticket to Ride, and the latest flavor of the month among those who hang out online at websites like BoardGameGeek. After paying the fee there's no pressure to hurry, just the temptation of the chilled craft beers, hot coffees, and snacks that are available to buy. And if you really loved the game you played, the Uncommons is ready to sell you a copy to take home.

May got the idea for the Uncommons after visiting Snakes & Lattes, a Toronto board game café that opened in August 2010. Snakes & Lattes is sometimes called North America's first board game café, but others such as the Haunted Game Café in Fort Collins, Colorado, predate it. While Snakes & Lattes might not be the first café, it certainly was the one that caught the media and board game community's imagination. So much so that in September 2015 Canada's public-access channel Fibe TV1 began broadcasting a sitcom about its early days called *Snakes & Lattes: The Show*.

Since Snakes & Lattes flung open its doors, board game cafés have sprouted throughout the world from San Francisco and Galveston, Texas, to Indianapolis and Salem, Massachusetts, and onto London, Paris, Berlin, and Sydney. The birthplace of the dedicated board game café, however, seems to be South Korea. By 2004 the Korean capital Seoul already had around 130 cafés renting out games and tables by the hour. Other East Asian countries followed suit and by 2012 there were around two hundred cafés in Beijing, and Singapore's The Mind Cafe had become an international chain thanks to the addition of an outlet in New Delhi.

While some of these cafés pop in and out of existence like mayflies, board game sales have been rising since 2013, suggesting that the trend is rooted in something more solid than hipster flights of fancy. In 2014 board game sales in the United States rose 9 percent over the previous year's, according to retail sales trackers NPD Group. In 2015 sales went up another 12 percent.

May believes the growing interest in tabletop games is, in part, down to Internet fatigue.

"I think there's been a definite pushback against the all-digital lifestyle," he says. "Even kids like a time to put down their phones, even for a few minutes at least, and ignore Twitter and Instagram and Snapchat.

"Also, geeky is popular, nerdy is popular. These things have come into the mainstream. So in the seventies playing Dungeons & Dragons may have gotten you into trouble with your parents and teachers and labeled you as very odd and made you an outcast. Now it's on TV and everyone seems to love those sorts of subcultures."

Board games, by their very nature, also bring us face-to-face with those we play with, he adds: "I love video games, but there's nothing like looking at your opponent across the table."

That personal, physical connection people have with board games is something May sees everyday from behind the counter of his store. "The greatest part of my day is when I can look out the windows and watch just about everybody who walks down the block stop, look in, and smile. Even if they don't come in you can tell they are thinking about it, you can tell that they are wondering and that they are curious."

The Washington Square Park game scene might be changing, but it's not dying. Instead it's evolving, just as board games themselves always have. And this small corner of Lower Manhattan is a microcosm of board gaming's place in the world. It not only showcases the all-ages, all-cultures appeal of board games, it also encapsulates their timelessness and resilience.

In the Chess Plaza people are playing a game so ancient that it was invented around the same time as matches and the concept of toilet paper. Even after the invention of the printing press, record players, cinema, radios, televisions, video games, and social media, we are still moving counters across boards and loving it.

But board games have done more than just survive.

They have made and ruined fortunes, revealed the secrets of lost civilizations and concealed the work of spies, and tested our morals. They have saved marriages, exposed the inner workings of our minds, decoded geopolitics, tracked societal changes, and organized the killing of millions. And—most of all—they have entertained us.

This book is the story of these board games. The games that shaped us, explained us, and molded the world we live in.

1

# TOMB RAIDERS AND THE LOST GAMES OF THE ANCIENTS

## *What board games reveal about our ancestors*

"At last have made a wonderful discovery in Valley," read the telegram. "A magnificent tomb with seals intact; recovered same for your arrival; congratulations."

A shiver of excitement shot through Lord Carnarvon's body. This was it, the message he had been waiting years for. It had been so long since he agreed to fund archaeologist Howard Carter's search that he had almost given up hope. After digesting the news, Lord Carnarvon ordered his servants to pack his cases. It was November 5, 1922, and Carter had just discovered the lost tomb of the ancient Egyptian pharaoh Tutankhamen. What awaited Lord Carnarvon in the pale sands of the Valley of the Kings was the greatest haul of Egyptian artifacts ever found. Grave robbers had ransacked most of the royal tombs millennia before, but Tutankhamen's burial chambers had barely been touched.

Carter and his team spent the next eight years clearing the site. For months workers scurried in and out of the tomb like ants, carrying relic after relic to the surface on canvas stretchers. Pots, shields, walking canes, stools, and fruit baskets mingled with beds of gilded wood, finely decorated chests, and thrones covered in precious stones and colored glass.

Among the treasures were four game boards. Some were plain and easily overlooked amid this archaeological bounty, but one stood out. The board sat on the upper surface of an oblong box that was held aloft by feline legs carved out of ebony. Paws at the bottom the legs rested on short golden drums attached to two sledge runners. The ivory veneer game board was divided into three rows of ten squares by a lattice of wooden strips, and at the front of the box was a drawer containing playing pieces and the short, flat throwing sticks that acted as the game's dice.

While the board uncovered in Tutankhamen's tomb was exceptionally ornate, Carter and his fellow Egyptologists had encountered this game before. They called it "the game of thirty squares," and it had been turning up in digs ever since explorers began excavating the ruins of ancient Egypt around the dawn of the nineteenth century.

The earliest confirmed set uncovered by archaeologists dates back to 3000 BC, around the same time that ancient Egypt was founded, but fragments of what appear to be the game's board have also been found in burial sites predating the kingdom's creation by centuries—discoveries that suggested the game could be as old as writing itself. Equally impressive, other excavations had found that the game was still being played when Alexander the Great conquered Egypt more than three thousand years after those first boards were made.

The game's presence in the ruins of ancient Egypt did not end with boards covered with centuries of dust. In many of their digs Egyptologists also discovered paintings of people playing the game on tomb walls. One such painting in Meir showed two players boasting to each other about how they would win the game. Some things, it seems, never change.

Even beyond the tombs the game lurked, its board scratched onto temple floors and carved into the wood of a quay on the River Nile. It also appeared in a satirical papyrus found in the ruins of Deir el-Medina, the ancient village that housed the workers who built the royal tombs in the Valley of the Kings. In the comic book–like papyrus a randy lion defeats a gazelle at the game and claims the chance to bed the antelope as his prize.

But some finds hinted at something stranger, something darker, about this game. In the tomb of Nefertari, the wife of Ramesses II, a painting showed the queen playing alone against an invisible opponent. The game also featured in the Book of the Dead created for the Theban scribe Ani around 1250 BC. In the book Ani and his wife are shown playing the game together as their bird-bodied spirits stand upon their sarcophagi.

Clearly there was something odd about this ancient game but for all the finds it remained shrouded in mystery. It resembled no game played in modern times, and for all their searching archaeologists found no record of the rules.

The tomb paintings revealed little. The side-on viewpoints of ancient Egyptian art obscured the position of pieces on the board making it impossible to decode the game from the images. The only secret the paintings gave away was that the ancient Egyptians called the game "senet," which meant passing.

The variety of boards added to the confusion. Half had all blank squares but the rest had hieroglyphs on the five squares in the bottom right of the board. Some boards had even more hieroglyphs; on one set every space was decorated. The playing pieces varied too. Some were cones, others were shaped like chess pawns and cotton spools.

With the rules unknown, Egyptologists could only speculate about the nature of the game but it was challenging even to reach a verdict on which square was the starting space. Some believed the game began at the bottom-right corner because the hieroglyph seen in that space on some boards meant "door." Others countered that this symbol also meant "exit" and argued that this space marked not the beginning but the end of the game.

The riddle of senet proved so vexing that in 1946 the Metropolitan Museum of Art in New York resorted to calling in George Parker, the founder of leading game manufacturer Parker Brothers. After examining the game and consulting with the museum's Egyptologists, Parker proposed a set of rules that he later used for a commercial version of senet. But his ideas were no more convincing than anyone else's. It was all just guesswork: modern ideas superimposed on ancient relics.

But as more and more senet boards piled up in museums across the world, a pattern emerged. The oldest boards tended to have blank squares but the most recent were decorated with religious hieroglyphs. Could it be that senet started as a game but later became an object of faith? The paintings of the game supported this idea. Early art showed the game as part of daily life but later artwork pictured senet in rituals and burials. Based on this and other evidence, the Egyptologist Peter Piccione proposed that over the centuries, senet morphed from game into a playable guide to the afterlife.

The ancient Egyptians believed that when people died their souls would gather on the barge of the sun god Ra at sunset and then be taken on a nighttime journey through the underworld. Along the way the souls of sinners would be punished and destroyed while the spirits that remained on the barge at sunrise would unite with Ra and live forever. Piccione argued that the hieroglyphs on the later senet boards represented key moments on this journey of souls. The top-left corner square of these later boards bore the symbol of Thoth, the ibis-headed deity who announced the arrival of the newly deceased in the underworld, and so this was where the game began. In the middle row of the board were squares representing Osiris, the green-skinned judge of souls who would send the guilty to be obliterated in flames, and the House of Netting where the impure would be entangled in nets and tortured. The last five squares on the board included the House of Rejuvenation, the mummification workshop where bodies were prepared for burial and eternal life, and the Waters of Chaos in which sinful souls would drown. The final space represented Re-Horakhty, the god of the rising sun, and signified the moment when worthy souls would join Ra for eternity. In this interpretation senet was no mere game but a gateway to the spirit realm. Through ritualistic playing of the game the living could learn what awaited them in the afterlife, and if fiery annihilation was to be their fate they could then change their ways.

The evidence collected by Piccione also suggested that the game's powers didn't stop there. Senet also acted like a Bronze Age Ouija board that allowed people to connect with the dead. They could even play senet against their own souls, which would explain that strange painting of queen Nefertari playing solo.

Ancient Egyptians were not alone in using games for fortune-telling. In the same year that Carter discovered Tutankhamen's tomb, another British archaeologist named Leonard Woolley began excavating the ruins of Ur in southern Iraq.

Founded circa 4000 BC, Ur became one of the richest and most populated cities of the ancient world. But its glory days did not last. Nomadic invaders sacked the city. The Tigris and Euphrates rivers that made Ur rich deposited enough silt to shift the coastline farther and farther from the city. War and drought followed, causing citizens to head for more fertile and secure lands. By the end of the sixth century the once great city of Ur was empty and slowly being buried by the shifting sands.

During his excavations of the city's royal cemetery, Woolley uncovered a game that became known as "the royal game of Ur," although other boards were later found across the Middle East. The game found in Ur had once belonged to a princess. Its squares were made from shell plaques inlaid into a wooden block and separated by intense blue strips of the precious gemstone lapis lazuli. Each square was decorated with intricate patterns: eyes, rosettes, and geometric motifs colored with red limestone and even more lapis lazuli. The board was distinctive, its shape reminiscent of an unevenly loaded dumbbell. The left side had an area four squares wide and three squares deep, which was connected in the middle by a two-square bridge to the right side, which measured two spaces wide and three spaces deep. Like senet, the royal game of Ur was a game of the dead. It had fallen out of favor hundreds of years before Woolley found the princess's board, and its rules were unknown.

For decades it seemed as if the rules of the royal game of Ur would remain a mystery, too, but then in the early 1980s Irving Finkel of the British Museum decided to inspect a near-forgotten tablet lurking deep within the London museum's vast archive of ancient relics.

The tablet's journey to London began in 177 BC, when the Babylonian scribe Itti-Marduk-balatu took a slab of moist clay and a blunt reed and began etching words into it using cuneiform signs, the earliest known form of writing. At the time the city of Babylon was in serious decline. Citizens were fleeing on mass to escape the constant battles for control of the city that followed the death of Alexander the Great and, in the chaos, the scribe's tablet ended up lost in the sand-covered ruins of what used to be the world's largest city. The tablet remained buried there until a team of European archaeologists rescued it from the dirt in 1880 and sold it to the British Museum, which catalogued it and then filed it away. And there the tablet sat largely ignored for yet another century until Finkel, the museum's cuneiform expert, finally got around to looking at it.

After taking it out of storage Finkel—a man who could easily pass for Professor Dumbledore thanks to his grand white beard and thin-framed spectacles—turned over the tablet and saw a pattern that resembled the distinctive board of the royal game of Ur. Curiosity aroused, he began translating the ancient script and, to his delight, discovered that it explained how to play the game.

The royal game of Ur was a race game. Players competed to get their pieces from the left side of the board to the exit on the right side by rolling dice made from sheep knucklebones. But as well as being an amusement, it also told players their fortune. Each of the board's squares carried a vague prediction that wouldn't seem out of place in a fortune cookie or newspaper astrologist's column. "You will find a friend," offered one space. Others promised that players would become "powerful like a lion" or "draw fine beer."

Using games for spiritual guidance or to learn about the future might strike us as strange today, but it makes more sense when we understand that our brains have a serious aversion to the concept of randomness. Our brains look for patterns in the world around us and instinctively try to identify the causes behind those patterns. It's an immensely useful ability. If we're hiking in the woods and hear an unexpected rustle in the bushes we are more likely to imagine a bear caused it than a random gust of wind. That interpretation will almost always be wrong but mistaking a breeze for a bear is no big deal while mistaking a bear for the wind is a big deal. Our brains' habit of formulating connections between events not only aids survival but also helps us develop theories and ideas we can put to the test, paving the way for new discoveries and insights.

Useful as this clearly is, our subconscious connecting of the dots also causes us to attribute nonexistent meaning to random events. Even today, some five hundred years since the mathematical theory of probability was developed, our minds still rebel against randomness. We might feel the dice are working against us during a game, imagine a secret conspiracy caused a tragic accident, or conclude a homeopathic remedy cured our cold.

The "Madden Curse" is a good example of how our minds impose meaning on randomness. Some American football fans believe there is a jinx on players who appear on the cover of the annual Madden NFL video game. After all many of the NFL stars who graced the game's cover got injured or underperformed that season. Some believe the curse is real enough to even campaign against their favorite players being made Madden NFL cover stars.

Of course there is no connection. American football is a rough game and injuries are commonplace. Take any random group of NFL players and you will probably find a good chunk of them were injured or performed below par in any given season, regardless of whether they appeared on the cover of Madden NFL that year. But because of the way our minds work, people imagine a cause-and-effect relationship, and every time a Madden NFL cover star gets hurt it reinforces the belief in the curse.

So if we are still creating bogus connections about unrelated events today it shouldn't be surprising that ancient people believed there was more to the results of their dice rolls and stick throws than mere chance. Instead of seeing randomness, people saw the invisible hand of the spiritual realm. Landing on the Waters of Chaos in senet was no random event but a message from a god, a ghost, or even your own soul.

Yet, for all the mysticism surrounding them, neither senet nor the royal game of Ur endured. In the case of senet, religion was probably its undoing. Under the Romans the Egyptians converted to Christianity and so the game was cast aside like the old gods. The fate of the royal game of Ur is fuzzier. Some believe it evolved into backgammon. An alternative theory is that early forms of backgammon drew players away from the royal game of Ur until it was forgotten. Or at least until everyone thought it had been forgotten.

For while the royal game of Ur died out in the Middle East, it lingered on unnoticed in the southern Indian city of Kochi. Sometime before the game died out in the Middle East, a group of Jewish merchants left the region and began an epic five-thousand-mile journey that eventually ended with them settling in Kochi. One of the things those adventurous traders took on their travels was the royal game of Ur, and their descendants were still playing a recognizable version of it when they began migrating to Israel after the Second World War, many hundreds of years after people stopped playing it in the Middle East.

The royal game of Ur is not the only board game that allows us to trace the footsteps of our ancestors and few games do this better than the mancala games of Africa, the Middle East, and southern Asia. Although widespread around the world, mancala games are less well known in Western countries, where they are sometimes portrayed as a single game, even though that's like calling playing cards a game. There are hundreds of different mancala games but what they all have in common is that they are two-player games in which people move playing pieces around a board of pit-like holes.

The most widespread mancala game is oware, which is also known as awari, awélé, and warri, among many other names. Oware boards consist of two rows of six pits. Each player owns the row nearest to them. The game starts with each pit filled with four counters or "seeds" that traditionally consisted of shells, nuts, or small pebbles. The aim is to capture the majority of these seeds. On his turn, each player chooses a hole on his side of the board, scoops up all the seeds inside, and then moves counterclockwise around the board, dropping one seed into each hole until his hand is empty. This process is known as "sowing." If the last seed sowed brings the number of seeds in a rival's hole to two or three, the player captures, or "harvests," all the seeds in that hole. And if the hole he sowed before it also has two or three seeds, he gets to harvest those seeds too, a process that continues until the player reaches the end of his opponents' row or encounters a hole that does not contain two or three seeds. The key to success is to sow seeds so that you harvest as many seeds as possible while limiting your opponent's ability to claim seeds from your side of the board.

Oware is straightforward but other mancala games are notoriously complex. One of these headache-inducing mancala games is bao, which is mainly played in East Africa. Bao's board features four rows of eight pits and comes with an intimidating list of rules that dictate the various ways to win, how the direction of sowing changes depending on the stage of the game, and how harvested seeds are to be redistributed on the board. Under certain conditions players must start to sow again from the hole they dropped their last seed into and, in theory, this chain reaction of sowing can be never ending.

In between the simplicity of oware and the complexity of bao are hundreds of strains of mancala from the three-row boards of Ethiopia and Eritrea to the twenty-four-hole version played by Roma Gypsies in Transylvania. And the distribution of these variants provides a breadcrumb trail of human migration and communication over the centuries.

The starkest evidence of this can be seen in how mancala games spread along slave trade routes. Oware, for example, came with the slaves taken from West Africa to the Caribbean, where they re-created the game's board in the soil. Much the same happened in East Africa under Omani rule in the seventeenth century. The slaves taken by the Omanis from Mozambique to Muscat brought with them a four-rowed mancala called njomba that they played in their homelands. Njomba spread from the slaves to the Omanis, who call it hawalis and still play it today. The Omanis also sold slaves to French colonists in the Seychelles, which led to njomba taking root there under the name "makonn."

But the bread-crumb trail of mancala games is a patchy one. People often played mancala games on makeshift boards scooped out of the earth or on wooden boards that rotted away. The generic nature of the seeds used in the game also makes it difficult to accurately trace the game's past.

The slippery archaeological record of the mancala games leaves much unanswered. We don't know if Africa or the Middle East was the birthplace of mancala games, or when the first games originated: we can only narrow it down to sometime between 3000 and 1000 BC. Nor do we know how the evolutionary tree of mancala games fits together. It could be that simpler games like oware came first and then grew into more complex creations like bao, but for all we know mancala games could have been dumbed down over time.

But there is one, much more recent board game whose evolution and spread around the world is far better documented: A game still being played today. A game molded by centuries of migration, war, trade, technological development, and cultural change. And that game is chess.

# 2

# CHESS: THE "MAD QUEEN'S GAME"

## *How the chessboard came to embody centuries of world history*

It's doubtful that many who frequent Washington Square Park's Chess Plaza today have ever heard of the Gupta Empire, but if it wasn't for this nearly forgotten Indian realm none of them would be there.

In the fourth century the Guptas ruled the lands around Pataliputra, a settlement near the present-day city of Patna in northeastern India. It was an unremarkable realm, just one of the hundreds of minor-league kingdoms that governed the Indian subcontinent at the time. But the Guptas were destined for greatness. In 319 AD the ambitious Chandragupta I became the kingdom's raja and everything changed. Chandragupta I wanted more than the small realm he had inherited, so he ordered his regiments of foot soldiers, chain mail–clad horsemen, and fearsome war elephants to build him an empire. Many local rulers took one look at his powerful military and surrendered without a fight. Those who resisted were soon crushed.

By the end of the century, the Guptas controlled most of India. Their empire stretched all the way from Balkh in Afghanistan to the mighty Brahmaputra River in Bangladesh. Its northern frontier snaked along the foothills of the Himalayas and in the south its lands extended down to Mumbai on the subcontinent's western coast and the Krishna River in the east.

To maintain this vast empire, the Guptas built an enormous army boasting more than half a million soldiers and a navy of more than one thousand ships. Yet, the Guptas were anything but oppressive despots, for as they expanded their territory the Gupta rajas ushered in one of the greatest golden ages in human history.

Under the Guptas, India flourished. Doctors developed new surgical techniques. The streets hummed to the sound of new musical instruments. Art, poetry, and literature blossomed. The science of metallurgy made huge strides and the Gupta rajas used their immense wealth to build schools, hospitals, and orphanages.

The empire's astronomers examined the stars and figured out that the earth spins on its axis and rotates around the sun, centuries before the rest of the world reached the same conclusion. Most significantly of all, the empire developed decimal mathematics and the concept of zero. And, before it disintegrated in sixth century due to inept rule and foreign attacks, the Gupta Empire gave the world chess.

Chess evolved out of an ancient Indian board game called ashtāpada. Players rolled the dice and hoped to be the first to get their playing pieces to complete a circuit of the board, which consisted of sixty-four squares arranged into eight rows and eight columns.

Sometime in the fifth century, however, people began using the ashtāpada board for a new four-player game called chaturanga, which means "four limbs" in Sanskrit. Chaturanga was a war game and the playing pieces represented the four divisions of the Gupta Empire's impressive military: infantry, horsemen, war elephants, and ships. Each player also had a piece that represented the raja who commanded their forces.

The infantry moved like chess pawns, always marching forward and capturing diagonally. The horsemen leaped around in the same distinctive L-shapes of chess knights and the war elephants charged vertically and horizontally along the board like rooks. And, just like the king in chess, the raja could move one square in any direction. The ships, however, were unlike anything from chess as we know it today.

They moved two squares diagonally in any direction and could also hop over any piece that blocked its path.

With its varied playing pieces, chaturanga must have struck people as something special. At the time, tokens in other games usually stood for the player on the board, but chaturanga's miniature re-creation of the Gupta military represented not the player but the army the players controlled. In chaturanga the players were gods and even the raja followed their orders.

Despite similarities to chess as played today, chaturanga was a very different game. For starters, there were four armies and the expectation was that players would work with an ally to defeat the other two players before turning on each other. There were also restrictions on how pieces captured each other that echoed India's caste system. Lowly infantry were barred from taking higher-ranking pieces and rajas could never be killed, only taken captive. Players who captured an enemy raja could even do prisoner swaps to get their rajas back in the game. Another big difference between chess and chaturanga was luck. Unlike modern chess, chaturanga was not a game of pure logic. Instead players flung dice-like throwing sticks into the air to find out which piece they could move. It was also usually played for stakes. There were even wild tales of players betting their fingers on the game and only walking away from the board once they had no more digits left to offer.

For the religious, chaturanga's status as a gambling game was a problem. The Hindu legal text the Laws of Manu opposed playing games with dice, and while its rules were treated as ideals to live by rather than laws to be enforced, it did deter the devout from playing. The Buddhists were even more disapproving and urged their followers to refrain from even learning the game. Under religious pressure, people got rid of the game's dice and gave the choice of which piece to move to the players.

Over the years more refinements followed. The war elephants and ships swapped their movement patterns and, most significantly of all, chaturanga became a two-player game. The shift to two players saw the once-allied armies of the original version unite into a single sixteen-piece force. To avoid having two rajas per army, one was demoted to a minister, an advisor to the raja who was limited to moving one square diagonally. And with only two rajas in play, the prisoner swaps no longer made sense, but since killing rajas was frowned upon, chaturanga became a game about trapping—checkmating—the opposing raja.

Merchants soon began taking this version of chaturanga with them on their travels along the silk roads, which provided not only a means for transporting goods but also games and ideas. One of the first places the traders brought chaturanga to was Persia. The Persians loved the game so much they even created a legend about its arrival.

The story begins with an Indian ambassador arriving at the court of Khosrow I, who ruled Persia's Sasanian Empire from AD 531 to 579. On meeting the Persian shah, the ambassador gave him a beautiful game board and a set of playing pieces fashioned from emeralds and rubies. As Khosrow I admired the elaborate game, the Indian issued a challenge on behalf of his king.

"If your wise men cannot figure out how to play this game it will show Persia is intellectually inferior and so we will demand tribute and gold," said the ambassador. "But if your men do solve the riddle of this game then my king will acknowledge you as a deserving ruler."

The wise men of Persia spent three days examining the unfamiliar game until one declared that he had discovered how to play it. After the sage had defeated the ambassador at the game twelve times in a row, the Indians acknowledged Khosrow I as the rightful ruler of Persia.

The Persians not only keenly embraced the game they called "shatrang"—they improved it. They introduced fixed starting positions for the playing pieces and changed the rajas into shahs. The Persians also began warning each other whenever they threatened their opponent's shah and concluding games by declaring "shat mat"—the king is defeated. These practices endure today as "check" and "checkmate." The Indian minister became a farzin, the personal bodyguard of the shah, and since the Persians lacked a significant navy the ships were replaced with rokhs, meaning "war chariots," even though armies had not used these units for centuries. In short the Persians adapted the Indian game to mirror their own society, and it wasn't long after the game's arrival in Persia before it would be altered yet again.

In the early seventh century the Sasanian court became embroiled in more intrigue and bloodshed than a *Game of Thrones* box set. Would-be rulers executed their brothers, suffocated a queen with a pillow in her bed, started civil wars, died of plague, and slaughtered an eight-year-old king in their pursuit of power. While the Sasanians were busy stabbing each other in the back, the followers of a new religion called Islam were busy preparing to invade Persia. Having just conquered Arabia, the Muslim armies surged into Persia in 636 and, aided by the turmoil within the Sasanian court, conquered the whole empire in just fourteen years.

Just as trade brought chess to Persia, the invasion led to the spread of chess in the Arab world as the victorious Muslim armies took the Persian game back home. The game became a favorite of caliphs and scholars but in this theocratic empire it was inevitable that the burning question of whether Allah approved soon arose. The Koran warned against playing for lots but was chess a gambling game? Not always, but often enough to raise suspicion among the pious. Others fretted that chess was so absorbing it would distract people from their religious duties.

Chess had supporters though. Many thought it was just a game, and even Umar ibn al-Khattab, the caliph who ordered the invasion of Persia, defended it by declaring: "There is nothing wrong in it. It has to do with war."

The debate didn't stop with the act of playing chess. The playing pieces were another bone of religious contention. The Persians, like the Indians before them, used playing pieces that resembled the military units they were meant to be, but the Islamic Caliphate opposed the creation of images of humans and animals.

Eventually a consensus was reached: Chess could be played in private so long as it wasn't for stakes and the playing pieces did not mimic living things. As a result the game's figurines morphed into abstract shapes—rectangular blocks, tall cylinders, and curved cones—that had to be distinguished by height and shape alone. Once more chess had evolved to reflect the society it had arrived in, changing like a chameleon to fit in with its surroundings.

Not that these alterations ended Islam's arguments about chess. Following the Iranian Revolution in 1979 chess was banned, only to be unbanned in 1988. When the Taliban seized control of much of Afghanistan in the 1990s, they also outlawed the game and began arresting anyone caught playing it. And in 2014 when Saudi Arabia's most senior cleric, the Grand Mufti Abdul-Aziz Al ash-Sheikh, was asked for his opinion on chess on his weekly TV show *With His Eminence the Mufti,* he declared it forbidden.

After traveling through a mix of trade and war from India to Arabia, chess finally reached Europe in 711 when a Muslim army crossed the Strait of Gibraltar, landed in Spain, and set about conquering the Iberian Peninsula. Within nine years the invaders had seized control of most of Iberia, which they declared was now the emirate of Al-Andalus. The Islamic invasion terrified Europe, striking at its identity as a Christian continent, and for the next seven hundred years European armies fought back, eroding away the emirate until it was no more.

The arrival of Muslim power in Europe did more than provoke shock and war, however. It also led to a period of cultural exchange between the Islamic and Christian worlds. In the process Europeans were reacquainted with the writings of the Greeks and Romans, introduced to new mathematical concepts—and taught how to play chess.

Europe took to chess with zeal. It was in France by 760 and being played by Swiss monks before the end of the tenth century. By 1050 the game was in southern Germany, and before the twelfth century was over it had reached Scandinavia and the remote Isle of Lewis off the coast of Scotland.

As it traveled through Europe, chess wove itself into the daily lives of the nobility. Pages who aspired to become knights had to learn the game and chess masters joined jesters and musicians as entertainers of royal courts.

Chess also became a game of love. The medieval practice of having noble women constantly accompanied by chaperones meant there were few opportunities for courting couples to get some privacy, but it was acceptable for men to visit women in their chambers alone in order to play chess. Unsurprisingly chess became very popular with unmarried couples and the game featured heavily in the romantic poetry of the time, including in some versions of *Tristan and Iseult,* where the Cornish knight and the Irish princess fall in love while playing the game.

Unlike Islam's religious leaders, the Catholic Church saw the divine on the chessboard. One religious treatise on morality, attributed to Pope Innocent III but probably not written by him, said: "The condition of the game is, that one piece takes another; and when the game is finished, they are all deposited together, like man in the same place. Neither is there any difference between the king and the poor pawn: for it often happens that when the pieces are thrown promiscuously into the bag, the king lies at the bottom; as some of the great will find themselves after their transit from this world to the next."

The Dominican monk Jacobus de Cessolis continued the theme in what became one of medieval Europe's bestselling books, *The Book of the Morals of Men and the Duties of Nobles and Commoners–or, On the Game of Chess.* In the book the northern Italian monk depicts chess as a model of feudal society with all people, from the highest king to the lowest peasant, represented. Each person, he argued, must know their place and adhere to the restrictions of that rank, just as each chess piece is bound by the rules of the game.

Despite being united in its love of chess, Europe couldn't agree on *how* to play it. Every kingdom seemed to have its own take on the game. In Germany some pawns could move two squares on their first move instead of the usual one. In northern Italy the king could leap over pieces on his first move. In England players could choose between a short and a long version of the game, each with different starting positions for the pieces.

Even the identity of the playing pieces was contested. Since Europeans were unfamiliar with elephants, people decided to change this unit to something else, but what it became depended on where you lived. In France the elephant became *le fou*—the fool or jester. The Germans disagreed and declared that the elephant should become a messenger. No, no, countered the Italians, it's a standard-bearer. In fact, you're all wrong, interjected the English, because it clearly should be a bishop.

But by the late 1300s, what Europeans all agreed on was that chess could be, well, a bit tedious. The core of the game had barely altered since the days when the Sasanian Empire ruled Persia, and for all its merits chess was a slow, plodding game. Armies would crawl slowly toward the middle of the board as if marching through mud. Chess was so slow that in the Middle East people would make multiple moves on their first go so that they didn't have to endure the grind of getting armies close enough to actually attack each other.

The Europeans wanted more action, more speed, and more aggression. And since chess was a folk game, something passed down through generations and owned by nobody at all just like traditional songs, people began trying to fix it. Some enlarged the board but that only made playing the game take even longer. Others rearranged the starting locations of each piece in the hope that would help. It didn't. In the Spanish kingdom of Castile, players unknowingly took chess back to its roots by letting dice decide what piece a player could move. For a while dice chess, which was both faster and easier to play, thrived, but the randomness robbed the game of its strategic depth. And once the novelty wore off, the dice joined the elephants, shahs, and ships on the scrapheap of chess for a second time.

Europe's chess experiments also led to the creation of checkered boards that made it easier to follow the game and, around 1100, someone in southern France repurposed the disc-like counters from backgammon and a chessboard to invent a new game that after several centuries of refinement became checkers.

The eureka moment in Europe's quest to jazz up chess came when people homed in on the idea of making the playing pieces more powerful. In hindsight this solution is obvious: if a playing piece moves too slowly around the board, why not let that piece move faster?

So the bishop or fool, or whatever it happened to be, lost its ability to hop over opponents but could now travel any distance diagonally. The pawns got an upgrade too. The two-square opening moves played in Germany became the norm and, as a countermeasure, the "en passant" rule was introduced. En passant, which is French for "in passing," is a special capture move for pawns that only applies under certain conditions. Let's say the white player moves a pawn for the first time, and moves it two squares so that it is adjacent to a black pawn. The black player can then, on his next go only, capture the white pawn using en passant by moving his pawn one square diagonally so that it is directly behind the white pawn.

The most radical change of all, however, was the rise of the queen as the most powerful piece on the chessboard. When chess first came to Europe, it inherited the Arabian vizier piece that was still, beneath it all, the same puny adviser to the raja conceived in India and limited to moving just one space diagonally. Much like the elephant, the concept of viziers didn't resonate in Europe so people called it the queen. But this was just a change of name; the moves remained the same.

In the fifteenth and sixteenth centuries, however, the queen morphed into the most powerful unit on the board, able to move any number of squares in any direction. The empowerment of the queen was not just about adding action to the game. It also reflected how female leaders were taking charge in kingdoms across Europe and showing they could rule as well as any man.

Leading the new wave of female rulers was Isabella I of Castile, who ruled alongside her husband Ferdinand II of Aragon as an equal partner. Together the couple reunited Spain, began building an empire in the New World, founded the Spanish Inquisition, and in 1492 conquered Granada, the last fragment of Muslim territory in Iberia.

While Isabella I was transforming Spain, another bold woman was making waves in Italy. Caterina Sforza believed her husband, the Count of Forlì, lacked what it took to rule a kingdom, and so she took matters into her own hands. Knowing that the death of Pope Sixtus IV in 1484 could lead to a new pope who would strip her family of their territory, Sforza went on the offensive. Despite being seven months' pregnant, she got on her horse and thundered toward Rome at high speed. On completing the ten-mile gallop, she occupied the papal fortress of Castel Sant'Angelo and refused to give it back until the new pope had guaranteed her family's land and title. She got what she wanted.

Yet some people still didn't get the message that Sforza was not a woman to be messed with. In 1488 a band of conspirators murdered her husband and took Sforza and her children prisoner. The rebels then demanded that she order Forlì's garrison to surrender. She persuaded her none-too-bright captors that she would need time inside the garrison to negotiate the surrender and on being released she ordered the troops to prepare to defend the city. After it became clear Sforza had tricked them, the rebels threatened to execute her children in front of her as she watched from the battlements.

According to one account Sforza responded by lifting up her skirt to expose her crotch and screaming, "I don't care, look, I can make more." Other reports say she merely gave them the Renaissance equivalent of the middle finger. Either way Sforza conveyed very clearly to the rebels that she wasn't going to budge. The shocked rebels didn't know how to respond and failed to carry out their threat. Soon after Sforza crushed the rebellion, rescued her children, and celebrated by dragging the conspiracy's ringleader around the streets behind her horse before having him disemboweled alive in the town square.

No wonder the Italians called the new version of chess "the mad queen's game."

More and more female leaders would follow from England's

Elizabeth I to Catherine de' Médici, queen to King Henry II of France, each helping to shatter the illusion that only men could govern. In this context, it's no surprise that players also came to accept the idea that the queen could be the most powerful chess piece.

The addition of the super-powered queen transformed chess. Suddenly the slow-motion battles of old became aggressive and snappy. Checkmate could even happen in just two moves. The queen made chess deeper too, adding another layer of complexity to the game and opening up new strategic possibilities. This wasn't a mere tweak but a transformative moment for the entire game. The game that had begun all those centuries ago in India was now almost indistinguishable from the game we play today.

The empowered queen spread fast. In 1536 the new queen was still a novelty in German-speaking kingdoms but eighty years later few people could recall the old rules. The feeble queens of old were quickly forgotten across Europe.

Chess continued to move with the times through the eighteenth and nineteenth centuries, but this time it was the players, not the rules, that changed.

As the Industrial Revolution spawned a new wealthy middle class, interest in chess increased. The newly affluent saw chess as a game of distinction, a pursuit worthy of refined minds. Chess had a particular appeal for the individuals who were ushering in the Age of Enlightenment, with its admiration of logic and reason.

Few encapsulated the game's appeal to that era's intellectuals better than the English writer Mortimer Collins, who in 1874 wrote: "There are two classes of men: those who are content to yield to circumstances, and who play whist; and those who aim to control circumstances, and who play chess."

The coffeehouses frequented by the intelligentsia responded to the interest in chess by opening clubs where boards could be rented by the hour. These chess clubs became places where players could watch the experts play, find worthy opponents, learn the game, and meet like-minded people.

Like precursors to the board game cafés of today, chess clubs opened across Europe during the eighteenth century from Berlin and Budapest to Edinburgh and Moscow. None, however, was more renowned than the Café de la Régence in Paris. Located close to the Louvre, it attracted legendary chess players and world-changing figures alike. Voltaire and Jean-Jacques Rousseau played there as did a young French lieutenant named Napoléon Bonaparte. The chess-loving Benjamin Franklin visited the club while living in France and it was here that Karl Marx first met Friedrich Engels.

The Café de la Régence might have had the big name clientele but London's chess scene also had many great players, and after decades of distant rivalry the British and French clubs began holding contests in 1830s to sort out who was best (the French initially, in case you were wondering). These tournaments marked the start of international chess competitions, and in the years that followed those initial Anglo-French showdowns, more and more chess clubs began hosting matches. And as competitions became more common, expert players became more serious about chess and invested more time in studying past games and developing new strategies to outwit their rivals.

But as players left their homes to play in unfamiliar clubs they found that not everyone played chess in the same way. The British chess player George Walker, who played at the Café de la Régence in 1839, complained that Frenchmen constantly offered their opinions on in-progress games. "They do not hesitate to whisper their opinions freely, to point with their hands over the board, to foretell the probable future, to vituperate the past," he moaned. "I have all but vowed that when next I play chess in Paris, it shall be in a barricaded room."

The Prussian chess master Adolf Anderssen, meanwhile, found playing against the English a chore. They sit "straight as a poker," repeatedly scanning the board, and only move after their opponents have "sighed hundreds of times," he grumbled.

These clashes of national character paled in comparison to a more fundamental problem world-traveling chess players faced: working out what any of the chess pieces in front of them were supposed to be. Chess might have existed for more than a thousand years but there was still no consensus as to what its pieces should look like. Some sets were literal representations of soldiers, other sets maintained the legacy of Islam's influence with militaries made from abstract shapes. Many sets fell between the two extremes, offering pieces that were symbolic, such as knights reduced to horse heads.

Soon players began talking about how the world really needed a standardized chess set. The chess clubs tried pushing their own designs onto the world. The Café de la Régence produced a set but its beanpole pieces were prone to falling over, and since height was the main distinguishing feature of the pieces, it was easy to get muddled about whether the piece with the thimble-like head was a bishop or a knight. The London chess club St. George's also had its own set, but while its sturdy bases meant the pieces rarely fell over, the button-like rings that formed the bodies of the pieces obscured the player's view of the board. What international chess needed were pieces designed with global tournaments in mind, and in 1849 the London game manufacturer John Jaques & Son delivered the design players needed: the Staunton chess set.

The Staunton set is what we now think of when we think of chess. Ask people to draw an everyday chess pawn and they will sketch the ball-headed one from the Staunton set. When we think of the default chess bishop with its miter headdress, we're thinking about the Staunton design.

The set is named after the British player Howard Staunton. At the time the set was created he was the nearest thing the world had to a chess champion but he did no more than endorse the design. The man who patented the set was Nathaniel Cooke, editor of the *Illustrated London News,* for which Staunton wrote the chess column, but some think the manufacturer's proprietor John Jaques himself may have been the real designer because of his abilities as a master turner. Regardless, the Staunton pieces struck the right balance between elegance and practicality.

Inspired by the neoclassical architecture that was in vogue at the time, the bases and columns of the pieces echoed the classical balusters that architects were incorporating into the balconies and staircases of new buildings. The knight continued the neoclassical theme by echoing the straining horse that pulls the chariot of the moon-goddess Selene in the Elgin Marbles, which had been removed from the Parthenon in Greece and placed in the British Museum in 1816.

The miter-wearing bishop, the horse-head knight, the castle-turret rook, the coronet for the queen, and the crown for the king had appeared in other sets but they had never been done with such attention to practicality. Each piece was thin enough to avoid obscuring the board yet also steady enough not to tip over easily. The individual pieces were instantly recognizable too. No one was going to get a Staunton bishop and queen mixed up. Just as crucially, the Staunton design wasn't too elaborate, making it cheap to mass-produce, which reduced the price of chess sets and allowed even more people to start playing the game.

The Staunton set was simple, unfussy, and elegant, and the emerging competitive chess scene embraced it. In 1851, two years after the set's introduction, the first truly international chess tournament was held in London. This tournament soon became a regular event, and as competitive chess grew, so did the Staunton set's popularity. And in 1924 the world chess federation FIDE cemented the set's position by declaring it the standard design for all international tournaments.

The professionalization of chess also ensured that the rules became the same the world over. Local variations disappeared and the European version of chess became the norm everywhere, including in the game's birthplace of India.

The folk game born in the Gupta Empire had traveled the world and back again for centuries. Along the way it had been molded by religion, war, female monarchs, and neoclassical architecture, and turned into an international sport. It had even played Cupid to courting couples.

But now the game was frozen in time and the same the world over. The game being played in Washington Square Park in 2016 is indistinguishable from the one that was being played by people in 1916, and it would be no surprise if chess was still the same in 2116 too.

3

# BACKGAMMON: THE FAVORED GAME OF INTERNATIONAL PACE-SETTERS AND ANCIENT EMPERORS

## *How backgammon became the most glamorous game of the 1970s*

It's the early 1960s and in the swankiest clubs of California a strange man is asking a strange question.

"Do you know how to play backgammon?" enquires the tall, gaunt man in a deep baritone voice. The conversations that follow his query are much the same wherever he goes.

"Backgammon, you say? The one with those shuttlecocks, like tennis?"

"No, not badminton—backgammon!" exclaims the man, his thick eyebrows rising up his forehead. "It's a game," he tells the baffled club members, "played on the board with those long, narrow triangles that you often see on the back of checkerboards."

"Oh that, I've seen that, but I have no idea how you play it, no idea at all," they inevitably respond.

"It's a fantastic game," says the man. "The very best game there is. A game of action. A game of skill. A game of fate. Let me teach you."

The strange man banging the drum for backgammon was Prince Alexis Obolensky, a Russian nobleman who had made it his mission to spread the word about the game.

That he even has to clarify the difference between backgammon and badminton is surprising. Backgammon has, after all, endured for centuries. It has seen dynasties come and go, and empires rise and fall. As a possible descendant of the royal game of Ur, it was already old when Christianity and Islam were founded. World wars, epidemics, droughts, urbanization, industrialization, and the space race—backgammon had been through it all and survived.

In the Middle East, where it is known as nard, backgammon is infused into the cultural DNA. It is played on the streets, in the coffeehouses, on beaches, and at home. Wherever there are people there is backgammon. It is said that in Iran every home must have three possessions: a copy of the Koran, the collected works of the great Persian poet Hafez, and a backgammon set.

"In Iran everybody knows how to play from childhood on," says Chiva Tafazzoli, the Iranian-born president of the World Backgammon Association. "In these countries, the children are born with the game under their arm."

Backgammon is a simple game with hidden depths. To understand it requires decoding that unusual board with the narrow triangles. These triangles are called points and are the spaces that the game's checkers use to move around the board. In total there are twenty-four triangles, divided equally between the two sides of the board. The twelve triangles on each side are split down the middle to form two zones, each of which contains six points.

Let's assume you are playing as white. The bottom-right zone—the rightmost six points on your side of the board—is your inner board, the place you want to move your checkers to. Moving clockwise, the next zone is your outer board. Above that is black's outer board and to the right of that is black's inner board, the place your opponent wants to get their checkers to. The goal is to move your checkers around the board and into your own inner board. So as white you move your checkers counterclockwise around the board while the black player's pieces travel clockwise.

Movement is controlled by two dice, and the number rolled on each are treated separately. Let's say you rolled a two and a four. You could move one checker two spaces and another checker four spaces. Alternatively, you could move one checker four spaces and then move that same checker another two spaces. Deciding how to use your rolls of the dice is where skill comes into the game.

But there are restrictions on movement if a point contains your opponent's pieces. Points that have two or more of the other player's pieces on it are blocked and you cannot move there. However, if a point has just one of their checkers you can move there and remove it from the board. Your opponent then has to use one of their dice rolls to move that checker back into the game through your inner board.

The final stage of the game is the "bearing off," when checkers exit the game for good after reaching their respective player's inner board. To start bearing off all your checkers must be inside your inner board. Then, turn by turn, you remove checkers from the board depending on the rolls of the dice. So if, as the white player, you roll a four and a six, you can bear off checkers from the fourth and sixth point from the right. The player who bears off all their checkers first wins.

The game's appeal lurks in its near-even mix of skill and luck. Rolls of the dice are crucial but so is how you use those rolls. This balance has made backgammon popular with gamblers since time immemorial, and no backgammon gamble is more famous or bloodier than that recorded by the ancient Greek historian Plutarch.

With help from Greek mercenaries and encouragement from his mother, Parysatis, Cyrus the Younger attempted to topple his brother, the Persian king Artaxerxes II, only to be slain on the battlefield. Furious, Parysatis set about exacting revenge on everyone involved in her preferred son's death. One by one they fell as Parysatis found ways to convince Artaxerxes to have them brutally executed.

Soon just one remained: Masabates, the royal eunuch who cut off the head and right hand of Cyrus so that his death could be proven to the king. But Masabates was careful and avoided any action or words that could give Parysatis a pretext to have him sent to the executioners. So Parysatis turned to backgammon. She often played the game with the king and one day she challenged him to play for a thousand gold darics.

She deliberately lost the game, paid her debt, and then, pretending to be upset at the defeat, urged him to play again but this time for possession of a eunuch. The trap laid, Parysatis defeated the king and claimed Masabates as her prize. After having the unfortunate eunuch flayed alive, she had him impaled on three intersecting posts and his skin hung before him so he could see it as he died.

Like the Greeks, the Romans also imported an early version of the game we now call backgammon from the Middle East. Initially it was known as ludus duodecim scriptorum, "the game of twelve lines," but following some changes to the board it became known as tabula, the Latin for "board." Both versions were immensely popular with the Romans, many of whom played it for money. Boards for the game were found carved into the courtyards of many villas in the ruins of Pompeii, and there was even a painting depicting two men playing it in a tavern before getting into an argument and being kicked out by the innkeeper.

Rome's emperors were just as hooked on the game as their subjects. While, perhaps unsurprisingly, Caligula stands accused of being a cheat, his successor Claudius loved the game so much that he wrote a book about it and had a board affixed to his chariot so he could play on the move. Domitian, meanwhile, was said to be an expert player while Nero reportedly gambled away enormous sums of money playing the game.

The Roman legions took tabula to every corner of the empire, introducing the game to what is now southern Germany, France, and the Netherlands. The game even made it to the empire's northernmost frontier in the north of Britain, where the locals took to calling it tables. But when the Roman Empire disintegrated and the legionnaires retreated home, tables began to fade away. The former subjects of the Romans simply failed to embrace the game with the same vigor as their departed masters.

But then came the Crusades, and many of the Christians who went east to battle with Muslim forces discovered and fell in love with the game during their time in the Holy Land. The Christian soldiers played it so much during the Third Crusade that England's Richard I and King Philip II of France issued a joint decree that banned anyone under the rank of knight from playing games for money and capped the sums knights and clergymen could bet. Those who disobeyed faced being "whipped naked through the army for three days."

The Crusaders brought the game home with them, reviving European interest and making it a feature of life in the Middle Ages. From royalty and aristocracy to people of no rank who played in inns that provided their customers with boards, the game of tables became a common sight across Europe. Tables even got name-checked in medieval literature, with mentions in both the epic French poem *The Song of Roland* and Geoffrey Chaucer's *The Canterbury Tales*.

Not everyone was pleased by tables' popularity. While it approved of chess, the Catholic Church saw the spread of this gambling game as anything but welcome and spent much of the Middle Ages campaigning for it to be banned. The church scored some victories, even persuading King Louis IX to ban the game in France in 1254, but for the most part people ignored the clerical disapproval.

As with chess, the rules for tables were far from fixed across Europe. As the Middle Ages gave way to the Renaissance, Europeans were playing as many as twenty-five versions of the game, each broadly similar but all with minor differences. In France there was tric-trac, where scoring points by carrying out particular moves took precedence over the bearing off of checkers. In Sweden there was bräde, which offered players twelve ways to win the game, and in Britain there was Irish, which was much like modern backgammon bar some small differences in the bearing-off process.

Eventually, sometime before the 1640s, Irish sired yet another variation. This British creation's main additions were twofold. First, if players threw a double they could use the numbers rolled for four moves rather than two. Second, it added a scoring system aimed at gamblers, which gave players extra points depending on how far behind their opponent was when the game ended. The British called this new version "backgammon," a name derived from the words "back" and "game," and in 1743 Edmond Hoyle, the famous English chronicler of card and board games, codified the rules that largely endure today.

Backgammon became the preeminent version of the game in Europe. It spread from Europe to North America and remained popular throughout the eighteenth and nineteenth centuries. Thomas Jefferson played it while taking a break from writing the U.S. Constitution. Admiral John Jervis of the British Royal Navy complained that the surgeons of Horatio Nelson's fleet wasted too much time on the game, and Jane Austen's Emma played it with her hypochondriac father. So how was it that Prince Alexis Obolensky struggled to find anyone who recognized backgammon as he went from club to club in the 1960s?

Backgammon's troubles began early in the twentieth century. The stigma around playing cards had eased and gamblers began snubbing backgammon in favor of whist and bridge. By the middle of the 1920s it looked as if backgammon's time in the sun was, at least outside the Middle East, over.

Then as the 1920s drew to a close the game got an unexpected shot in the arm from a new addition to the rules: the doubling cube, a large six-sided die bearing the numbers two, four, eight, sixteen, thirty-two, and sixty-four. The doubling cube's job was to keep track of how many times the original bet had been doubled.

Let's say we're playing for a dollar. At the start of my second turn I decide—in a fit of overconfidence—that the game is going to go my way so I take the cube, turn it so that the two is face up and request a doubling of the stakes.

Now you have a dilemma. You can accept the stakes being doubled or concede the game. If you concede you have to pay me the original dollar stake, so you decide to accept the double—after all it's not much and the game has barely started. So now we're playing for two dollars and you have control of the doubling cube.

Later on the game starts going your way, so you decide to use the cube to double the stakes. Now I face the same choice you did. Do I fold and pay you two dollars or up the risk to four dollars? I go with the double and now it's me in charge of the cube and we're playing for four dollars.

This back and forth between us with the cube continues throughout the rest of our game. Sixty-four might be the highest number the cube can offer but since there's no cap on how much doubling that can go on in a game, we double all the way to one hundred and twenty-eight. Now we're playing for one hundred and twenty-eight dollars instead of one and both of us are praying that the dice will go our way.

The doubling cube transformed backgammon, making it as much a battle of nerves as a contest of skill and luck.

Yet who created the doubling cube is unknown. The best guess is that it emerged around 1925 in one of the exclusive American clubs where backgammon lingered on, but there are claims that it may have been used in Paris first. Whatever its origin, the cube revived interest in the game among wealthy gamblers, who found the threat and thrill of ever-spiraling stakes too alluring to ignore.

For a few years backgammon found itself back in vogue, but the doubling cube's refresh of the game didn't last, possibly stymied by the onset of the Great Depression. With the doubling cube unable to reverse the trend toward card games, backgammon's popularity faded. As the 1930s became the war-torn forties, interest in the game waned again and the decline only accelerated after the Second World War. By the late 1950s, backgammon was unloved and near forgotten in the West. Rarely played outside the older establishment clubs of the rich, it mainly lingered on—largely out of tradition—as the game on the back of checkerboards. Backgammon came to be viewed as old and tiresome, a game from a bygone era.

As the American humorist Robert Benchley wrote in 1930 at the peak of the doubling cube's reenergizing of the game: "I can remember a time when backgammon was something you played only when you had tonsillitis, and didn't think it was so hot even then. . . . In fact, backgammon was the spinach of indoor sports, and something that was reserved almost exclusively for little visiting cousins and children who 'weren't very well.'"

By the dawn of the 1960s backgammon's popularity had sunk so low that even being lampooned as "the spinach of indoor sports" would have been an improvement. But then Prince Alexis Obolensky came to the rescue.

Born in the imperial Russian capital of St. Petersburg in 1914, Obolensky came from aristocratic stock. His blue-blooded family were respected members of Tsarist Russian high society and descendants of the Rurik Dynasty that founded the medieval state of Kievan Rus', the spiritual predecessor to Russia.

Obolensky, however, never got to know the land of his birth. In 1917 the Russian Revolution brought the communist Bolsheviks to power and the Obolenskys fled the country. Like many White Russians, as the Russian émigrés who fled after the revolution came to be known, the Obolenskys wound up in Istanbul. And it was during his family's time there that Obolensky learned how to play backgammon from their Turkish gardener and developed his passion for the game.

By the time Obolensky was an adult his family had relocated to New York, where the young prince made a name for himself in Manhattan's Russian community as an enfant terrible thanks to his playboy lifestyle, womanizing, and taste for gambling. By the Second World War he was spending much of his time in Palm Beach dealing in real estate and working as a U.S. intelligence agent tasked with sniffing out German spies trying to infiltrate America from the south. After the war he returned to his playboy ways and began fighting a one-man battle to encourage his circle in the private clubs to take up his favorite game. But, as his trip around the clubs of California at the start of the 1960s proved, it was an uphill task. Much to his disappointment, even those who played backgammon rarely gambled big on the game.

Then he got a call from the recently opened Lucayan Beach Hotel in the Bahamas. Hoping to drum up interest among the wealthy, the owner of this multimillion-dollar resort hotel asked the well-connected playboy if he could persuade a group of rich backgammon players to come there and play for big money. Obolensky thumbed through his bulging contacts book and invited thirty-two wealthy players to take part in what he billed as the first International Backgammon Tournament.

In March 1964 Obolensky and his invited gaggle of counts, millionaires, stockbrokers, and socialites arrived in Freeport for a long weekend of backgammon and big bets. "The rest of the people in the hotel looked at us as if we were from Mars with all those checkerboards," Obolensky told his nephew Valerian years later.

The tournament proved to be great fun, a sun-kissed party lubricated by drinks on the house and sound-tracked by the constant rattle of dice shakers.

The final game saw New York publisher Porter Ijams up against Charles Wacker III, a millionaire racehorse breeder from Chicago who later became a fugitive after the FBI accused him of tax evasion worth millions of dollars. At stake was nearly eight thousand dollars, around sixty thousand dollars in today's money, and the silver "Obolensky Cup." Wacker won the game, and after he and those who bet on his success collected their winnings, the backgammon glitterati wound down with a black-tie celebration that featured games played for even greater stakes and concluded with a swim at sunrise.

Although the shebang was a publicity stunt for the hotel, by the time the party was over Obolensky understood how he could bring backgammon back from the dead. If these people would fly a thousand miles or more despite the risk of getting knocked out in the first round, then there had to be something in this backgammon tournament idea. As he flew back to Palm Beach, Obolensky was already plotting a second tournament.

The next year the tournament returned to the Bahamas, but this time with sixty-four players, including some who had come all the way from Europe.

In 1966 the event came to London, where it was hosted by two of the British capital's most exclusive clubs, Crockford's and the Clermont.

Like the New York Racquet and Tennis Club, these two Mayfair venues were among the last Western bastions of backgammon during the 1950s, places where the pre-WWII interest sparked by the doubling cube never went away.

Among the regulars on the London and New York backgammon club circuit was Lewis Deyong, who made his money in Manhattan real estate. The London scene was, he recalls, very much a behind-closed-doors affair but peppered with the kind of society figures who filled the gossip columns of British newspapers. People like James Goldsmith, the billionaire tycoon who inspired the corporate-raider character Sir Lawrence Wildman in Oliver Stone's 1987 movie *Wall Street,* and Lord Lucan, the British earl who would vanish without trace in 1974 after the murder of his children's nanny Sandra Rivett.

"There were a lot of people and a very active game there, high stakes," says Deyong. "But the Clermont's founder John Aspinall would never let any newspaper people through the front door, he had a horror of that. This was private."

The arrival of the International Backgammon Tournament in London widened the pool of jet-set players yet again, with Deyong, the Greek shipping magnate Aristotle Onassis, and James Bond film producer Albert Broccoli becoming familiar faces on the global backgammon scene.

With so many wealthy gamblers now hooked on backgammon, it was only a matter of time before Las Vegas called, and in 1967 Obolensky was asked to bring the tournament to Sin City by Sanford Waterman, the manager of the Sands Hotel casino who later moved to Caesars Palace, where he famously pulled a gun on Frank Sinatra during an argument about the singer's casino debts. Later that year the World Backgammon Tournament, as it was now being called, landed amid the neon, glitz, heat, and glamour of the Las Vegas Strip and it was bigger than ever. This time 128 players and their families signed up for another weekend of high rolling on the backgammon board. The winner was Tim Holland, a chain-smoking New Yorker, who got hooked on backgammon at a Miami country club where he used to play golf for bets in the 1950s. Holland said it took him several years and losses of around thirty thousand dollars to master the game, but after winning in 1967 and 1968, when the competition went to Las Vegas for a second time, his backgammon balance sheet was in the black.

While Holland celebrated, the defeated commiserated in the bars. After hearing one sob story too many, the tournament staff put a sign above their desk detailing their listening charges. One hard-luck story: $1. To listen with interest and sympathetic nods: $2.50 extra. To say, "That's the unluckiest game I have ever heard of": another $5. But rather than putting people off, the losses spurred them on. Thanks to the game's mix of chance and skill, winners could be dismissed as lucky while losers could blame their misfortune on the dice. Next time, they told themselves, the dice would be on their side.

The World Backgammon Tournament soon spawned a global network of regional and club championships that allowed the most avid players to spend their lives in a constant round of first-class flights and backgammon contests in plush hotels. From Monaco and Athens to Miami and Rio de Janeiro and onto London and Vienna, wherever the so-called beautiful people gathered there was backgammon.

The big spenders might have been hooked but the wider world was still only dimly aware of the snowballing backgammon scene. Knowing that backgammon needed something more to capture the world's attention, Obolensky teamed up with writer Ted James and penned the first English-language book about the game since 1950. Published in 1969, *Backgammon: The Action Game* promised to teach readers the ins and outs of "the favorite game of the international pace-setters." The self-aggrandizing blurb worked. The book sold hundreds of thousands of copies.

As the 1970s began, backgammon had shaken off its stale image. It was no longer the spinach of indoor sports but the foie gras, a game associated with wealth, luxury, and glamour. "I made people think they should be doing it, that only the best people were involved," Obolensky told *Time* magazine. "We brought in snobbism."

Backgammon became the game to be seen with. The list of movers and shakers who played or at least made a point of being seen around the backgammon tables reads like a who's who of the 1970s. There were Hollywood stars like Roger Moore, Joan Crawford, Michael Caine, and Polly Bergen. Sports personalities abounded, among them legendary Detroit Tigers' slugger Hank Greenberg and acclaimed NBA star Larry Bird, perm-haired English soccer pro Kevin Keegan and the world's number one tennis player Jimmy Connors. Deyong even remembers playing against members of the Dallas Cowboys. Hugh Hefner threw backgammon parties at the Playboy Mansion West in Los Angeles that attracted the likes of Diana Ross, Motown founder Berry Gordy Jr., supermodel Margaux Hemingway, Atlantic Records' supremo Ahmet Ertegun, Cher, Bond girl Jill St. John, and Linda Lovelace, the star of the notorious porn film *Deep Throat*.

"Hefner's house was the center of backgammon," says Deyong. "His house was just a mecca for all these famous types. Hefner loved backgammon in those days; we used to play until four in the morning. It wasn't a huge game we used to play. If you had a really bad evening you might lose one, one-and-a-half thousand dollars. Nobody's going to die for that."

Former Beatle Ringo Starr dropped in to watch the final games of the 1976 World Backgammon Championship in Monaco. The Rolling Stones' Mick Jagger and Bill Wyman were photographed playing the game on tour. Tina Turner even posed for the cover of *Las Vegas Backgammon Magazine*. Elsewhere backgammon was embraced by multimillionaire playboys like Gunter Sachs, major art dealers, Parisian property tycoons, and even Dr. Christiaan Barnard, the South African surgeon who performed the world's first successful heart transplant. The list goes on and on.

With so many jet-setters playing backgammon, the game became something everyone wanted to play. "Suddenly everybody had to have a backgammon board," remembers board game designer Mike Gray. "It got trendy and people started playing it. It was on the back of checkerboards everywhere forever but people never really played it, but it's a way better game than checkers."

Bloomingdale's doubled the amount of boards it stocked, F.A.O. Schwarz began selling as many backgammon boards as chess sets, and Abercrombie & Fitch dedicated an entire page of its 1974 Christmas catalog to the game. London department store Harrods said it couldn't get luxury sets in stock quickly enough to meet the demand.

Fashionable leather goods purveyor Mark Cross began selling luxurious sets that came in brown pigskin leather attaché cases and cost two hundred and eighty dollars. "Backgammon has taken off like wildfire," the company's fashion director told *Newsweek*. "It's *the* game to play."

Big business began throwing money at the game, hoping to get some of that backgammon sparkle to rub off on their brands. Soon the major tournaments were being sponsored by upmarket casinos, cigarettes, and liquors—from Philip Morris International Filters to Black & White whiskey—all brands that were constantly on the look out for ways to get around legal restrictions on their promotion. "Backgammon was very good for problem sponsors," says Deyong. "There were all kinds of different laws but nobody quite knew where to stick backgammon and nobody really cared that much."

It had taken Obolensky just ten years to realize his dream of reinvigorating backgammon. Yet by the middle of the 1970s, the big sponsors were growing tired of the man who had brought backgammon into vogue, not least because of the astronomical bar bills he would rack up at their expense. "The Obolenskys would sit around the hotel bar ordering champagne, it's ridiculous when you are on the cuff like that—you just don't do that, you go to an extreme not to," says Deyong. Eventually the sponsors sidelined Obolensky and asked Deyong to start organizing backgammon tournaments instead.

Deyong soon found himself running as many as twenty tournaments around the world every year and as the corporate sponsors inflated the prize pools, more and more people joined the tournaments, taking the money and excess to ever-greater heights. At the 1977 World Backgammon Championship in Monaco, as much as half a million dollars of prize money and side bets were at stake. By 1984 the opulence at the Monaco competition, which became the biggest backgammon contest of all, was even greater. Middle Eastern oil billionaires arrived on giant yachts, European playboys wearing Cartier watches arrived at the Monte Carlo Casino in white Cadillacs. Women dressed in high-end fashions would order drinks that cost a minimum of twenty dollars and competitors enjoyed an opening dinner prepared by world-class chefs.

When the competition started the casino halls became filled with the noise of dice shaking in cups and the hum of languages ranging from English, French, and Japanese to Hebrew, Farsi, and Spanish. In the background big spenders puffing expensive cigars would place bets of as much as fifty thousand dollars on individual games. And when the backgammon stopped, the party started. In one Monaco bistro a group of Mexican textile millionaires celebrated by throwing their wineglasses, crockery, and chairs into the fireplace and, once the fire had consumed all that, they began flinging fistfuls of money into the flames.

Yet sometimes the sums at stake were outrageous even by the standards of the wealthy. "I played one game for sixty-four thousand dollars—the biggest game I ever played," says Deyong. "That's probably about quarter of a million now and if you want to know the truth, yes it was nerve-racking."

The game was tight, the lead swinging back and forth between the players until they each had four checkers ready to bear off. Since it was his turn to roll, Deyong had the advantage. If he rolled a double the game was his. If he didn't his opponent would get one last chance to snatch victory by throwing a double. The odds against getting a double are five to one, enough to make Deyong the clear favorite. Even so, favorites do get beat. Everything now depended on luck and, to add to the pressure, the casino's owner had bet money on Deyong winning.

Since the math was on his side Deyong decided to double the stakes.

"I knew psychologically my opponent would take a doubling of the stake, so I said to this chap who owned the casino, I hope you won't mind if I double him to sixty-four thousand dollars and the guy said to me: 'You should be shot if you don't.'"

Deyong turned the doubling cube. His opponent accepted. Bet inflated, Deyong picked up the dice cup, began shaking and then let the dice tumble out. The fate of sixty-four thousand dollars was about to be decided. Everyone held their breath. "I was never very lucky playing backgammon, but for once the gods were with me and I threw a double two so he didn't get to roll, I just took four men off and that was it," says Deyong.

Such enormous bets became increasingly common following the 1973 oil crisis, which saw the price of gas quadruple in a matter of weeks as the Middle Eastern oil monopoly OPEC cut off supplies. The Arab oilmen who made mind-blogging sums of money out of the crisis soon began splashing their cash at the backgammon tables. But while the leading Western players weighed up the mathematical probabilities of success before making decisions about whether to accept or reject a doubling of stakes, the Middle Eastern players preferred their backgammon fast and fatalistic.

"They all thought they were the world's greatest players and this was the bonanza for the professional expert players," says Deyong. "They didn't have the doubling cube in the Middle East and they didn't realize that it was lethal. Their answer was *inshallah,* it's the will of Allah; that was their determining factor in taking a double. These people just watered the backgammon community for years."

As the money swirling around the backgammon set became a blizzard, the game began attracting professional bridge players who hoped their mathematical approach to game playing would make them big bucks. "These bridge players came from a different culture and it wasn't very entertaining to play with them," says Deyong. "The great bridge player Oswald Jacoby was entertaining to play with, but the rest of them would sit there staring at the board never saying anything."

"I actually played a bridge player in the world championship finals in seventy-three," he recalls. "Her name was Carol Crawford and she'd just sit there and look at every play forever. After about half an hour of this I hauled out a newspaper and started doing the *New York Times* crossword.

"Obolensky and Carol's husband John came over. Obolensky said: 'There are people from the press here, you're turning this into a farce.' I said: 'Look, it's not me, it's her. I can't just sit here for hours while she looks at an opening.' In the end they begged me not to do it."

Deyong found the contemplative pace of the serious players who were now flooding the tournaments frustrating and tried his best to hurry them. "I'd stand there and glare at them and if they didn't then make a move I'd say, 'You have ten seconds to make a move or you're going to get a penalty point,' and the guy would say: 'But why?' I'd say: 'You've got five seconds left.'"

Part of the problem was that unlike the playboys, the serious players moving into the game were there for the money. For them it was work, not playtime, and if taking forever to make a move psyched out their opponent all the better.

By the middle of the eighties the gulf between the professional players and the jet-setters began to corrode the backgammon scene. The wealthy playboys, tycoons, and heiresses found backgammon wasn't fun anymore and stopped coming. And when they went so did the glamour and the money.

"When I ran the big Monte Carlo tournament in our biggest year we had seven hundred people there. I understand that Monte Carlo in July 2015 had about one hundred and twenty," says Deyong. "That is what these genius game players achieved. They've driven out everybody else except themselves, and as a consequence backgammon has no interest to the wider public and they ruined what was a very entertaining kind of pastime."

Shorn of its glamour and money, backgammon lost its allure. Texas hold 'em poker became the gamblers' game of choice thanks to its simpler rules, TV appeal, greater number of players, and bigger prize pools. And the arrival of online gambling in the 1990s only hastened backgammon's decline. "Internet poker absorbed so much interest among people who liked to play games," says Deyong. "They would have quarter of a million people playing every night on virtual tables. Backgammon could never have done that because of the initially complicated nature of the game and the rules."

By the time Chiva Tafazzoli started playing in backgammon tournaments in the early 1990s, the scene was a shadow of its former self. "I'd heard of the glitz and the glamour of backgammon in the seventies and I found that these tournaments were nowhere close to being glamorous or nice or social," he says. "It's repelling more than attractive. There were very, very few people involved that you might want to put on the front of a glossy magazine."

In 2000 Tafazzoli began trying to give backgammon a makeover. "I thought maybe it is a good time to try and bring back the glitz and glamour to the game, a little bit at least, and created the World Backgammon Association," he says. "I started doing tournaments and supplied these tournaments with high-class boards. I gave back a standard to backgammon tournaments so that if there is a World Backgammon Association tournament you know what to expect. You expect a nice playing room, you expect to have nice classy venues, excellent equipment, good tournament direction, a lot of fun at the same time, and high-end competition."

Nonetheless it's a tough challenge, he admits. "In the old days you had a lot of people with a lot of ego and a lot of money coming to tournaments," he says. "Today you have empty pockets but cameras and computers on every tournament. People are recording their matches, they are doing the analyzing and it's basically the fewer errors you make the better you are. It has become a very, very technical game instead of a gambling game. This is the way it goes, this is what backgammon is today."

Despite losing its cachet, backgammon is still popular with professional poker players. "Some of the very big names of poker today have a backgammon background," says Tafazzoli. "Gus Hansen is still today a world-class backgammon player. Phil Laak is a backgammon player. Phillip Marmorstein is a backgammon world champion. These people have become professional poker players using the tools and the thinking process of backgammon for poker."

What these poker stars take from backgammon, Tafazzoli says, is an understanding of how to maximize chances and minimize risks. "The beauty of the game is that the better players will always win at backgammon in the long run, but the spice of the game is that nobody knows how long this long run is," he says. "On a short-term basis anyone can beat anyone, but in the long run the better player will always prevail. The world champion could lose a game or two to Stevie Wonder but in the long run the world champion will ultimately win against someone weaker than him."

And maybe the same is true of backgammon itself. Maybe in the long run the game of backgammon will prevail once more. Taking the long view, backgammon is a winner that has been dogged by a recent run of bad luck, but who can tell where the dice will fall next?

## 4

# THE GAME OF LIFE: A JOURNEY TO THE UNIQUELY AMERICAN DAY OF RECKONING

### *What the Game of Life tells us about the development of U.S. society*

The sun may have been shining on Springfield, Massachusetts, in the summer of 1860, but for Milton Bradley these were dark days.

The twenty-three-year-old had begun the year with high hopes. In January he bought a state-of-the-art color lithography press and opened a printing business above a soda fountain next to Court Square. Orders poured in. Only the press's habit of misaligning prints or smudging ink at the slightest drop of sweat could detract from his new venture's success. That and his employee Jack Keppy. Keppy could "pull a press when he wasn't drunk," Bradley wrote in his diary. Trouble was Keppy was always drunk.

Keppy didn't last long and neither did the initial rush of orders. By the summer a recession was looming and Bradley found himself with an idle press and no customers in sight. His mood darkened, and the nagging fear that he would end up destitute started to consume him. He put his wedding on indefinite hold to avoid subjecting his wife-to-be to a life of counting pennies, and his friends began worrying about Bradley's morose state of mind.

One concerned friend, George Tapley, attempted to lift Bradley's mood by suggesting they play a board game called the Mansion of

Happiness. First released in Britain in 1800, the Mansion of Happiness crossed the Atlantic in 1843 thanks to the publisher W. & S. B. Ives of Salem, Massachusetts.

The Mansion of Happiness offered a puritanical twist on "the game of the goose," a mindlessly simple game where players rolled dice and moved counterclockwise around a spiral track while being sent backward or forward by special spaces along the way. Created in Italy in the sixteenth century, the game of the goose and its many variants became a favorite among wealthy Europeans from the child king Louis XIII of France to thrill-skilling aristocrats who played it for stakes. But where the game of the goose promised amusement, the Mansion of Happiness offered to enhance the soul. To achieve this lofty goal, it took a hard line on sin, imposing harsh punishments on players unlucky enough to land on an evil-deed space.

"Whoever becomes a Sabbath Breaker must be taken to the Whipping Post and whipt," thundered the rules. "Whoever gets into a Passion must be taken to The Water and have a ducking to cool him." Elsewhere drunks were put in the stocks and perjurers locked in the pillory. Players who landed on Immodesty or Audacity would be sent back to the space they came from and instructed to "not even *think* of happiness, much less partake of it."

Not that thinking of or partaking in happiness was likely when playing the Mansion of Happiness; it's about as much fun as being in the dock at a Salem witch trial. But for Bradley, a devout Methodist Episcopal who grew up in a New England community that regarded games as sinful distractions from the godly activities of work and prayer, it was inspiring. Almost immediately he decided he should get into the games business and create a religious game of his own. As he recounted in an 1893 letter, "I happened to play this game one evening at the house of a friend, and conceived the idea of a game which a short time after was published by us as our first venture in the game business, and christened, 'Checkered Game of Life.'"

Acutely aware of how his pious peers might react to him making games, Bradley marketed the Checkered Game of Life as a thinly disguised Sunday school sermon billed as "a highly moral game . . . that encourages children to lead exemplary lives." The aim was to live a good life and become the first player to achieve a happy old age by amassing one hundred points while traveling around the sixty-four red and white squares on the game's checkerboard. While the red squares were blank, the white squares depicted the highs and lows of life as viewed through the moral prism of nineteenth-century America. There were squares of virtue that advanced players to spaces that rewarded them with points, and squares of vice that pushed players away from the fifty points available to players who landed on the "Happy Old Age" space in the top-right corner of the board.

The connections between these squares of vice and virtue revealed much about the Protestant mind-set of the time. Players who landed on Honesty advanced to Happiness, Perseverance led to Success, Bravery to Honor, and Industry to Wealth. The vice squares, meanwhile, warned of life's pitfalls. Gambling hurled players to Ruin, Idleness resulted in Disgrace, Intemperance led to Poverty, and Crime meant a spell in Prison. Darkest of all the vices was the Suicide square, which eliminated players from the game and carried a chilling illustration of a man dangling from a noose tied to a tree branch.

Some squares were neutral. Players who landed on Truth, Matrimony, or Jail faced no prizes or punishments. The same was true of Poverty, of which Bradley's rules said: "It is not necessarily a fact that poverty will be a disadvantage, so in the game it causes the player no loss to pass through poverty."

The game's underlying message was clear: everyone is responsible for his or her own actions and all setbacks—bar suicide—can be overcome if you strive for goodness. Disadvantage due to external factors like disease, race, or circumstances of birth are nowhere to be found within this game.

The Checkered Game of Life further reinforced its message of personal responsibility through the way players moved around the board. In line with its antigambling stance, the game rejected dice. Instead, players twirled a teetotum, a small spinning top with a hexagonal cardboard disc that bore the numbers one to six mounted onto a spindle. The teetotum performed exactly the same job as a die, but without the stench of gambling attached.

Unusually for the time, the numbers on the teetotum not only dictated how many spaces players could move but also the directions they could go. Spin a one and you could move up or down the board one square. Spin a five and you could move one or two squares to the left or the right. Sometimes the edge of the board forced your move, but most of the time players got to choose. Would they seek Happiness by moving to Honesty? Or would they head to School so they could go to College?

Next to the pious punishments of the Mansion of Happiness, the Checkered Game of Life looked like a flag bearer for political correctness, but Bradley's mix of chance and player agency meant his game made its moral point more effectively. The Mansion of Happiness left everything to chance. You spin the teetotum—again no devilish dice allowed—and found out if you were to be a saint or a sinner. Chastity or immodesty, it all came down to the hand of fate. But in Bradley's game the path to happiness depended on making the right choices and pressing on when faced with adversity.

Despite this moral underpinning, Bradley still fretted that his fellow believers would disapprove of his creation. Religious New Englanders viewed board games as gateway drugs that could lure children into a life of gambling and sin. Fearing a backlash, he decided that rather than selling his game to the people of Springfield, he would sell it in New York City.

The Big Apple seemed like a good place to sell a virtuous game. New Englanders viewed New York as a den of wickedness, a place where the teachings of Methodist and Puritan preachers fell on deaf ears. Bradley saw this as an opportunity. The lack of devotion, he reasoned, might make New Yorkers more open to—and possibly more in need of—his game, and if it sold well there, favorable word might spread beyond the city limits. So in September 1860 Bradley boarded a train to New York with a trunk filled with several hundred copies of the Checkered Game of Life. Within two days of arriving in the bustling metropolis, he had sold every copy he had to Manhattan's department stores, news vendors, and stationers.

It was a promising start but on returning to Springfield, Bradley lost interest in the game as he cashed in on the buzz surrounding Abraham Lincoln's bid to become president by selling color prints of a portrait of the man expected to become the next White House resident. The prints were an instant hit with the solidly antislavery people of Springfield and for a few weeks Bradley's terrible summer became a distant memory.

All was going swimmingly until an eleven-year-old girl from Westfield, New York, wrote Lincoln a letter in October 1860. The girl's name was Grace Bedell and she supported Lincoln but when her father came home with a picture of the presidential hopeful she was not impressed. So she grabbed her feather-quill pen, dipped it in the inkpot, and wrote Abe a letter. The crux of Grace's missive was blunt. Abe, grow a beard. Now. "You would look a great deal better for your face is so thin," she explained. "All the ladies like whiskers and they would tease their husbands to vote for you and then you would be President."

A few days later a doubtful Lincoln replied. "As to the whiskers, having never worn any, do you not think people would call it a piece of silly affection if I were to begin it now?" he asked. Despite his reservations, Lincoln took Grace's advice. He stopped shaving and began cultivating the stately mustache-less beard it is now hard to imagine him without.

After word spread of Abe's new look, demand for Bradley's Lincoln prints vanished and angry customers began asking for their money back. Deflated, Bradley hauled the thousands of now unwanted Lincoln portraits he had printed into the yard outside his office and set them on fire. Once again the young entrepreneur was facing ruin. But then the Checkered Game of Life came to the rescue.

Bradley's hunch about New York had been right. After finding favor in the Big Apple, word of mouth about his game began radiating out to New England and beyond. Orders began trickling in, building in number day by day until they became a flood. By spring 1861 Bradley had sold more than forty thousand copies of the Checkered Game of Life and reinvented himself as a toy and games maker. He never looked back.

The Milton Bradley Company went from strength to strength. It thrived during the Civil War by selling pocket-size versions of the Checkered Game of Life, backgammon, and chess to Union soldiers. Eventually the company became one of the biggest game manufacturers in the United States.

The Checkered Game of Life remained on sale for decades, but as the years passed interest waned. By the end of the nineteenth century the game seemed like a relic, a flashback to an era of hardscrabble rural living and quill pens that had been replaced by the skyscrapers, electric lights, and telephones of a country that had now become the world's largest economy.

Aided by industrial-scale printing technology and the growth of leisure time, U.S. board game manufacturers stopped issuing finger-wagging religious lessons to children and began selling entertainment.

Leading the move away from pious play was George Parker, who formed Parker Brothers in 1888. His game-making career began at his childhood home in Medford, Massachusetts, where he would often play games with his friends. One day a friend complained about the moral preaching that underpinned all their games and how he wished games were more fun. Rising to the challenge, Parker invented a card game called Banking, where players sought to get rich through speculation. His friends loved it and so the enterprising teenager began selling currants on the streets until he had enough money to print copies of the game to sell.

By the 1890s Parker Brothers and other publishers were releasing games that were all about having fun and being modern. Instead of threatening players with trips to the whipping post, board games now gave people a taste of the new cycling craze with the Game of Bicycle Race, let them rise from errand boy to chief executive in the Game of Telegraph Boy, or embark on an exotic adventure with To the North Pole by Air Ship. Players could even enjoy tabletop shopping trips with the Game of Playing Department Store, which tapped into the excitement about the department stores that were opening in cities across the nation.

Even ancient morality-minded games like the Indian game moksha patamu moved with the times. In the game, which drew on Jain and Hindu beliefs, players would travel the board toward spiritual liberation. Along the way they would climb ladders that represented virtues such as knowledge and generosity and slide down snakes that represented vices such as anger, vanity, lust, and killing. With its oversupply of snakes compared to ladders, moksha patamu's message was that the path of goodness is harder than the path of evil. But when moksha patamu reached Victorian Britain in the 1890s as Snakes & Ladders, the vices and virtues had been removed, and there were as many ladders as snakes resulting in a message-free plaything. And when the game eventually reached the United States in 1934 as Chutes & Ladders, even the snakes were gone.

The aging Bradley was unimpressed with these new games. In an 1896 article for *Good Housekeeping* magazine he harrumphed about how "healthful fireside" games like the Checkered Game of Life had been sidelined by novelty games in "gaudy" and "showy" boxes.

Few listened to Bradley's objections. The Checkered Game of Life's time was up, and eventually it disappeared from the Milton Bradley Company catalogs. But as its hundredth anniversary approached, the game came back from the dead. Only this time it was reinvented for the TV generation in a way that probably had Bradley turning in his grave.

Reuben Klamer was the man behind the game's resurrection and in 1959 he was big news in the toy business. "I was sizzling hot," he says, grin widening and eyes lighting up at the memory. Klamer had struck toy gold the year before with the hula hoop. "Every child in

Australia had a hula hoop, although it wasn't called a hula hoop then," he says. "The fad was so great in Australia I thought I could duplicate that in the United States."

With help from the television star Art Linkletter, Klamer launched the Spin-A-Hoop. Together with Wham-O's Hula-Hoop, the Spin-A-Hoop ignited the biggest toy craze of the 1950s. By the end of 1958 millions of American kids were gyrating their hips to keep these plastic hoops spinning and Klamer had become a sought-after toy inventor for hire with a plush Beverly Hills office to match. The challenge now was how to follow a smash hit like the Spin-A-Hoop. Klamer soon hit on an idea. "I looked in *Toy and Hobby World* magazine and I saw an ad by the Milton Bradley Company for crayons," he says. "They manufactured crayons and that gave me an idea to go to them and try to sell them the idea of a children's art center."

In June 1959 Klamer flew to Springfield, Massachusetts, to present his art center vision to Milton Bradley president James Shea Sr. Shea wasn't interested but asked Klamer if he would have a go at creating a board game to mark the company's 1960 centenary. "I'll try," replied Klamer. "Will you let me look through your archives tomorrow morning?"

The next morning Klamer found himself rooting around the darkest corners of the Milton Bradley Company archives. And there, amid the dust and clutter of one hundred years of toy making, he found a box for an unfamiliar board game: the Checkered Game of Life. "The word 'life' struck me, electrified me," he says. "I thought, '*That's* a theme for a game.' I didn't see the game. I didn't know how to play it, but the name Life is what got me."

Klamer kept quiet about his discovery, said his good-byes and then spent the flight home scribbling in his notebook—thrashing out ideas, dreaming up concepts, and sketching plans for a game about life. Once back in California he teamed up with artist Bill Markham and got to work on a prototype. Keen to make the game stand out on television, Klamer used his plastics know-how to create three-dimensional buildings and scenery to give the game visual impact. He also created small pink and blue plastic pegs that would represent the members of each player's tabletop family and get slotted into the holes of the tiny plastic station wagons that acted as the playing pieces.

Two months later, Klamer was back in Springfield to show the Milton Bradley Company the result: the Game of Life. The company loved it and before September 1959 was over he had a deal.

Like the Checkered Game of Life, Klamer's creation reflected its time, capturing the optimism and consumerism of white America at the dawn of the 1960s. But instead of the checkerboard used in Bradley's 1860 game, the Game of Life turned the journey through life into a long, winding road that snaked around the board, passing plastic buildings and cutting through polyethylene mountains. The road of life sometimes forked but the final destination was always the same: the Day of Reckoning, the moment when the value of players' tabletop lives would be measured by toting up their money to see who was richest. Good deeds were out, greenbacks were in. One hundred years on from Bradley's morality lesson, players no longer aspired to a happy old age but dreamed of spending their retirement in the luxurious surrounds of Millionaire Acres.

And with money the key to victory, the most important choice in the game came right at the start, when players decided whether to go to college or head down the path of business—a choice that determined how much they would earn whenever they passed one of the many Pay Day spaces on the board. Those who went into business would earn five thousand dollars every Pay Day while those who opted for the more expensive college route would get the chance to land a high-flying career as a doctor, lawyer, physicist, teacher, or journalist. And, this being 1960, all jobs were for life. As the rules stated: "Once a doctor, always a doctor."

After that early decision, choice vanished almost entirely. Instead, the rest of life's journey depended on spinning the large rainbow-colored Wheel of Fortune that sat upon a plastic mountain in the middle of the board. Like the Checkered Game of Life's teetotum, this eye-catching and satisfying-to-spin wheel was born of prudishness about dice. "I was told by Mr. Shea: 'Do not use the dice, dice is an evil, and besides Monopoly uses dice and we don't want to imitate Monopoly.'" Klamer laughs. "The spinner came about as a result."

Although the legacy of Puritanism explained the use of the spinner, religion barely featured in Klamer's game. Aside from a compulsory stop at the church to get married (no bachelors, spinsters, or living in sin allowed), religion is absent. The only god in this American life was money.

The unwritten message may have been that money is what sorts life's winners from losers, but there were other lessons for players too. It is almost always wiser to go to college than not. If you don't buy insurance you run the risk of crippling bills should you crash your car, or if a cyclone rips your home to shreds. You can borrow from the bank, but you will have to pay it all back with plenty of interest on top. Players could also invest in stocks—a novel concept for a time when just one in ten U.S. citizens were stockholders, compared to around half today. But while you could make a fortune betting on the stock market, you could lose your shirt too. Klamer saw the inclusion of stocks, insurance, and debt as educational. "Children didn't know what stocks were, so that was an opportunity for parents to tell children about stocks," he says.

For some the lessons in the Game of Life were life changing. One life the game changed was that of the heart surgeon who later saved Klamer's life. "When I was young and we used to play the Game of Life, every time I was the doctor in the game I would win the game, and that started me thinking that maybe I ought to be a doctor," the surgeon told Klamer on discovering that his patient had created the game.

The Game of Life's mirroring of reality proved to be a winning formula. When Milton Bradley unveiled the game at the New York Toy Fair in February 1960, it was an instant smash with retailers. "The immediate interest was evident," says Klamer. "At that time the big buyers would carry their order pads with them and they placed big orders months ahead of production."

Backed by big orders, an endorsement from Art Linkletter, and one of the first national television advertising campaigns for a board game, the Game of Life became a runaway success. Today, it has been in production for more than fifty-five years and its various editions and spin-offs, which include My Little Pony– and *Star Wars*–themed incarnations, have sold tens of millions of copies worldwide. And its popularity shows no sign of fading, with the game racking up its highest-ever single-month sales in its history in December 2014. Klamer credits the game's enduring success to its basis in reality: "It's down to earth. Everyday information was in that game, it represented life. People related to it, people were into having babies on the way and insurance and all the things in life that were within it."

But the game on sale today is not the same game that wowed retailers at the 1960 New York Toy Fair. The Game of Life is a game in flux, constantly moving with the times, and its alterations reveal much about how U.S. society has changed since it first reached store shelves.

Look closely at the 1960 edition and you'll soon find spaces that clash with our present-day outlook. Take the space where players are cheerfully informed that they've discovered a uranium deposit worth one hundred thousand dollars. Encountering a crop of radioactive ore might not feature high on life's wish lists today, but in 1960 many Americans dreamed of just such a find. Encouraged by the U.S. Atomic Energy Commission's promise of big bucks for discovering uranium deposits, fifties America went uranium crazy. Thousands of amateur prospectors armed themselves with Geiger counters bought from Sears catalogs and guides like *From Rags to Riches with Uranium,* and headed into the southwestern deserts in search of an atomic jackpot. The excitement even inspired a board game, 1955's "educator-approved" Uranium Rush, where players searched for uranium using a battery-powered Geiger counter. The real-life uranium rush would eventually fizzle out in the mid-sixties after the Atomic Energy Commission withdrew its cash incentives.

Today, the chance to find uranium in the Game of Life is gone. It is even absent from the recent reproduction of the 1960 edition, where it was replaced by the less alarming discovery of a "mineral deposit." Also missing from that reproduction is the space requiring players to buy a coat made from raccoon fur. Now it's just a coat—no fur required.

The decision to airbrush these less palatable aspects from the remake of the original came from Hasbro, the game's current owners. "Hasbro sometime ago changed those to be socially in tune with the times and I'm not sure how much of a fight we put up, but they would not entertain us going back," says Philip E. Orbanes, cofounder of Winning Moves, the company that reissued the 1960 edition.

Raccoon-fur coats and uranium mines were not the only spaces from the original 1960 version of the Game of Life that would be phased out. Another was the Poor Farm, the final resting place for the game's losers. Introduced in the 1880s, poor farms were tax-supported institutions that took in people who could not pay their own way. The impoverished elderly, the bankrupt, the homeless, the insane, the orphaned, and the disabled—they all wound up on the poor farms. In return for a bed and regular meals they would, if able, work the land until the Grim Reaper came for them, at which point they would be buried on the grounds in unmarked graves. By 1960 poor farms were on their way out thanks to the introduction of Social Security, but the terrifying prospect of winding up there still lingered in the collective consciousness.

As in reality, the poor farm offered a depressing end to the Game of Life. If players reached the Day of Reckoning without enough money to win, their only hope was to gamble everything they had on a single spin of the Wheel of Fortune in the hope of being declared a Millionaire Tycoon and instantly winning the game. And if their lucky number didn't come up, to the poor farm they went.

The poor farm didn't stay in the game for long. But its initial replacement wasn't much sunnier. Instead, players who fell short of Millionaire Acres now ended the game "destitute and disgraced" and "reduced to living on Social Security" in a small house.

By the end of the 1970s, however, that dismal destination morphed into a tempting Country Cottage following another rethink of how the game's Day of Reckoning worked. The new ending offered players a choice. If players thought they had more money than everyone else they could move into the Millionaire's Mansion to sip champagne and live off the profits from the yachts, racehorses, private jets, and other status symbols they had bought on their way around the board. Alternatively, players could hibernate in the Country Cottage earning a pension equal to their salary every turn while waiting for the remaining players to finish.

Klamer's original "winner takes it all" capitalism had been tempered. Now, even if you weren't the richest player you still got to enjoy a pleasant retirement. The change was probably made to offer all players a happier ending, but it was also in sync with the politics of the Reagan presidency that followed a few years later in the 1980s. President Reagan's economic policy rested on the belief that everyone in society benefits from lower barriers to wealth and job creation, something that the Game of Life's new conclusion echoed. You might not get to spend your old age in Millionaire's Mansion living out Duran Duran MTV fantasies with the help of a luxury yacht, but you could still enjoy a successful life. Everyone was a winner now, just maybe not the biggest winner.

The 1980s version of the game also tweaked how players' cardboard lives began. Instead of choosing whether to buy auto insurance before heading down the road of life, every player now started the game with vehicle coverage. This change reflected the revolution in auto insurance that took place during the 1970s, as state after state made vehicle coverage mandatory. The oil price shocks of the 1970s also left their mark, with the gas-guzzling plastic station wagons from the 1960 edition swapped for more economical minivans.

While these changes fitted the Reagan years well, by the 1990s the Game of Life already seemed out of step with the times. The nineties were being billed as the "sharing, caring" decade—a fresh start after a decade of yuppies idolizing Gordon Gekko, stock market booms and busts, power suits, and excessive use of hairspray. In came Nirvana, out went Mötley Crüe. Good-bye President George H. W. Bush, hello saxophone-tooting President Bill Clinton.

In this climate the game's wealth obsession looked dated and Hasbro, which gained control of the game after buying Milton Bradley in 1984, knew it. Hasbro's answer was to abandon money altogether. The company envisaged a new direction for the Game of Life, one where good deeds took precedence over earnings. Instead of money, players would now collect Life Tiles that represented the experiences of a life lived well. Whoever ended the game with the most Life Tiles would be crowned the victor.

But the players whom Hasbro tested the revised game on hated it. They wanted the money back in the game. "They did try alternatives only to find that it seemed very important for people to win the money," says Beatrice Pardo, executive director of the Reuben Klamer Toylab. "So when they tried to change it to maybe what we would call politically correct alternatives it didn't go through. We found that money was the motivating factor, so it stayed."

The Life Tiles eventually reached the stores in diluted form in the 1991 edition of the Game of Life, which tried to find a middle way between the raw appeal of getting rich and the nagging sense that life should be about more than money. As before, players traveled the rocky road of life, but now they not only squirreled away their salaries, they also collected Life Tile cards whenever they landed on a space representing a socially minded deed, like saying no to drugs, helping the homeless, or recycling the trash.

Each Life Tile card carried news of a positive life experience. Players might discover they had written the Great American Novel, won a Nobel Prize, or opened a health food chain. But when it came to the Day of Reckoning these achievements were converted into cold, hard cash. Collect a humanitarian award, get one hundred thousand dollars. Compose a symphony, get a hundred and fifty thousand dollars. Saved an endangered species? Help yourself to another two hundred thousand bucks. Life, it seemed, always came down to money no matter what you did.

The 1991 edition also updated the career opportunities. College was now more expensive, landing players with thirty thousand dollars of debt rather than the barely noticeable dent in their finances that it once had. And instead of having your profession determined by where you landed on the board, you now drew your occupation randomly from a deck of career cards. You could wind up a doctor or an accountant, an athlete or a cop, or even become a "superstar."

The most telling change to the careers, however, was that no player had a job-for-life anymore. In the real world, the era of lifetime jobs seemed to be fading fast as globalization and a spate of corporate downsizing saw the American workplace become an increasingly insecure place for many. In a nod to this, the Game of Life's road was now filled with chances for career changes. Maybe you'd get fired, your career cut short in another round of downsizing, or find yourself having a midlife crisis that prompts you to quit your job in a fit of fear about getting old (where's a sports-car, status-symbol card when you need it?). Alternatively you could head to Night School in search of more lucrative employment. No matter how your career had ended though, there was always a new job waiting for you. The Game of Life was just too cheery to condemn anyone to long-term unemployment.

But while Hasbro shied away from the dark side of life, one man had taken the task of injecting more realism into the game into his own hands. That man was Chris Pender, who had spent much of his sixties childhood in Macedonia, Ohio, playing the Game of Life. "It was my favorite game," he recalls. "I really liked the spaces where something happened to you, like a tornado blew over your house or you inherited a skunk farm. Monopoly people were much more into negotiation and money. I was much more into Life because things are happening—you are playing a character almost."

But as Pender entered his teenage years and footage of the Vietnam War flashed across the family TV, he began to question the game's omissions. "When I was a kid Vietnam was going on, and I remember the night we all watched the draft lottery, which was televised and like the lotto," he says. "I remember we were all watching and my three older sisters were all worried about their boyfriends being drafted, but I was still pretty little so I was still playing Life, and that's when I first thought that would be an interesting space to put in the game: 'You're drafted and have to go to Vietnam.'"

As his teenage years continued the game's omissions only stacked up higher. Where was sex? What about race and gender? Why does the game pay scant attention to religion and life's small moments of simple joy?

Most people would have just moved on, accepting the reality that the Game of Life is designed for parents to play with their kids, so having little Johnny sent to war and little Joanna getting pregnant at seventeen would hardly encourage sales. But not Pender. His ideas for making the game more real never went away, and as ideas for improvements entered his head he recorded them in notebooks. One day, he promised himself, I'll make my version of the game.

Eventually, Pender got around to making his version: the Game of Real Life, a parody of his childhood favorite that ventured where Hasbro wouldn't dare. "It was like a midlife crisis," he says. "I turned forty and I felt this mortality, that I'm actually going to die, and thought I better get that game done before I die. So that's what motivated me to actually start working on it. The main parody is that happiness wins the game instead of money and I think that's the sociological statement. I'm not that interested in money as much as happiness and in my game you make choices based on that, and I think that's more where our culture is going, at least some of it."

Released in 1998, the game was a ramshackle ramble through life where the goal was to become the happiest player on the board in a world filled with curveballs before heading off toward the white light at the end of the tunnel. Not that every player would make it that far, because in Pender's game players could die before old age. Their first brush with death came right at the start, where if they rolled a one they end up aborted and have to start a brand-new life. Later they could die in a mugging or—in a nod to the 1960 version of the Game of Life—after getting radiation poisoning from a uranium deposit.

Elsewhere players could find themselves dropping acid, living during wartime, or in need of an organ transplant. They could find God, end up prostituting themselves on the streets, experience snow for the very first time, wind up pregnant after having sex, go to the movies with their grandchildren, or get forced to take part in a corporate fitness program.

Each copy of Pender's game was made by hand and sold at the Portland Saturday Market craft fair in Oregon and, later, over the Internet. Despite this lo-fi production, it became a cult success. Since its debut in 1988 Pender has made and sold more than twenty thousand copies, enough for him to eke out a small living from it.

The Game of Real Life even became a favorite of educators working with troubled teenagers and drug and alcohol programs. "I think they just like the way it works when you land on a sex or drugs space and the drug or sex is offered to you," says Pender. "That was my experience. I remember I was at a party in my twenties and there was a mirror with white powder on it coming my way. When it got to me I remember I asked the person what it was and they said: 'I don't know.' I said, 'Oh,' so I just passed it along because if you don't know what it is why are you snorting it?

"So in the game you're in that kind of social situation where drugs are is being offered, and it's your choice whether you take the drug or not, so it gets into that responsibility thing. I think the Alcoholics Anonymous people like that whole 'It's your responsibility, not anyone else's' thing."

The topics in the Game of Real Life were never going to make it into the "real" Game of Life, but Hasbro didn't give up trying to make the game less about wealth. In 2007 it attempted to move away from money again with a spin-off called the Game of Life: Twists & Turns. Promising "a thousand ways to live your life," Twists & Turns omitted many of the Game of Life's familiar traits. Now there was no end of the road, just a looping network of paths leading in and out of four zones: Earn it, Learn it, Live it, and Love it.

Players could roam wherever they liked on the board. If they wanted their cardboard life to be about money, they could enter the Earn it zone, but if a happy family was desired, they could frolic in the Love it quarter. Or, if unable to choose, they could shuttle back and forth between zones, trying a little bit of everything.

Along the way players picked up Life Event cards that detailed more than a hundred experiences for their pretend lives. Maybe they would start a stop-smoking club at college while in the Learn it zone or sing the national anthem to the President in the Live it zone. In Earn it they could add an award-winning website to their résumé, or stop a burglar from robbing their neighbor in the Love it zone.

The big prizes were the life points gained from these experiences. For while the game still involved earning money, life points determined who won. At the game's end, which could take as many or as few turns as players wanted, everyone's money, experiences, and possessions were converted into Life Points by the game's Electronic Lifepod using a secret formula. The Electronic Lifepod also took on the role of the game's banker, recording the progress of every player with the help of the Visa-branded credit cards that players used instead of paper money.

The adoption of credit cards was timely. Twists & Turns arrived in stores at the peak of the oughties' credit boom and it shows. The Lifepod was a generous lender—the kind of bank that would have gone belly up or required a taxpayer bailout when the credit bubble popped soon after. In Twists & Turns you could be a stable cleaner making five thousand dollars a year, but if you wanted to live in a million-dollar mansion why let money get in the way? The Lifepod has the subprime mortgage for you. And just as in the real-life credit binge, players were free to borrow without having to think through how they would pay the money back.

Yet for all the choice and freedom it offered, Twists & Turns felt aimless. Empty even. It was life without meaning. Without the Day of Reckoning looming in the distance, all that was left to do was to circle the board until time ran out, floating from one random event to another like a jellyfish buffeted by ocean currents.

Twists & Turns might have escaped the financial goal that remains the basis for the main editions of the game, but in doing so it had revealed the Game of Life's most telling lesson yet: that a journey through life without a direction is just no fun at all.

5

# THE FORGOTTEN MESSAGE OF MONOPOLY

## *How Monopoly went from anti-landlord tirade to celebration of cutthroat capitalism*

The militiamen were trapped. Outside thousands of furious men, women, and children wanted them dead.

The militia had come to Pittsburgh to restore order, but when they began bayoneting and firing at the crowd of rebellious railroad workers and their supporters, the rioters did not run. Instead they began shooting back and hurling stones, bricks, paving slabs, and other debris until the troops were forced to barricade themselves in the roundhouse of the city's railroad terminus.

As the overwhelmed militia sought refuge, the crowd began setting fire to the terminus. Soon dense clouds of black smoke, illuminated by the glow of more than a thousand burning buildings and freight cars, filled the night sky. Then the mob commandeered an oil train, set it ablaze, and sent it rattling down the tracks and into the sand house next to the roundhouse. The flaming locomotive crashed into the sand house, setting it on fire. Soon thick plumes of acrid smoke began filling the roundhouse. Choking and spluttering, the militiamen grabbed every fire hydrant they could and began battling to hold back the flames.

By seven in the morning the militia decided they had no choice but to shoot their way out. As they left the mob surged toward them and gunfire from both sides rang out. By the time the militia finally escaped twenty people were dead and many more wounded.

The violence in Pittsburgh on the night of July 21, 1877, was no isolated incident. Across the United States striking railroad employees and citizens had brought the railroad network to standstill, paralyzing the economy. From Newark to San Francisco, Chicago to Virginia, trains stood idle and railroad company property burned. The violence would only end in September after President Rutherford B. Hayes ordered federal troops to go from city to city and reestablish order at gunpoint.

The Baltimore and Ohio Railroad's decision to slash their workers' pay by 10 percent was the spark behind this national explosion of anger, but the causes ran much deeper than the actions of one company.

The United States of 1877 was a nation divided. Most employees worked punishing hours in dire conditions for abysmal pay while the men they worked for amassed almost unimaginable fortunes. Men like John D. Rockefeller, whose Standard Oil company was busy building a monopoly on oil refining, and the railroad king Cornelius Vanderbilt. Men who in the decades that followed would only increase their stranglehold on American industry. No one embodied these Gilded Age tycoons more than J. P. Morgan, the financier who used his wealth and connections to create industrial Goliaths like U.S. Steel, the first billion-dollar corporation. Morgan embodied the popular image of the nineteenth-century capitalist: a cane-carrying banker with a white handlebar mustache who dressed in a tuxedo and top hat.

Some saw these men as captains of industry who were making Americans wealthier and lauded their philanthropy. Others branded them robber barons who used their power and riches to snuff out competition, exploit workers, and undermine democracy by bribing corrupt politicians. As these industrialists and financiers amassed millions, the cracks in society split open. People began talking about class struggle and forming trade unions to take the power back.

The economist Henry George believed he had a better answer to the inequality tearing the country apart, and in 1879 he set out his plan in a book called *Progress and Poverty*. George argued that undeveloped land was God given and any increase in its value was due to the work done by people. As such, the money landlords made simply from owning land really belonged to everyone and the government should take all of that money back on behalf of society by imposing a land value tax. The income from this tax, George believed, would be so vast that all other taxes could then be abolished, a move that would let people keep all the proceeds from their own labor. This, he concluded, would narrow the gap between rich and poor by encouraging more productive use of land, raising the income of workers, and preventing landowners from parasitically accumulating wealth without having to do anything to earn it.

Opponents countered that the plan would see people taxed regardless of their ability to pay, asserting that land cannot be made instantly productive, and that his tax might reduce rather than encourage investment, but George's arguments struck a chord all the same. *Progress and Poverty* became a bestseller and spawned a new political movement, the single taxers, who wanted to see George's theory put into practice.

Elizabeth Magie was one of those who joined the fight for a single-tax society. Born in Macomb, Illinois, in 1866, Magie was a modern-minded woman trapped in an un-modern age, an independent spirit who sought to make her own way rather than rely on a husband to provide for her. She worked, wrote poems about love and unfairness, penned short stories, and impressed her theatrical friends with convincing portrayals of boy characters. Magie was also an inventor. Aged twenty-six, she created a device that allowed paper to be fed into typewriters more easily and got it patented at a time when fewer than 1 percent of patents were owned by women. Her introduction to George's theories came when her father gave her a copy of *Progress and Poverty*. After reading it she became a single-tax disciple and joined the movement, eventually becoming secretary of the Women's Single Tax Club of Washington, D.C. When George died in October 1897, Magie, like many single taxers, vowed to keep fighting for his ideals, but without its charismatic leader the movement quickly lost momentum.

Undeterred, Magie searched for ways to revive interest in a single-tax system. She tried giving talks on the subject but felt she was reaching too few people. So in 1902 she designed a board game that would bring George's arguments to life by demonstrating the harm monopolistic landlords cause and how a land value tax was the cure.

She called it the Landlord's Game.

In the game players traveled around and around the board using paper money to buy lots, railroads, and utilities. After buying a property players could then charge rent to anyone who landed on it and build houses that increased the amount they could demand. Each time players completed the circuit of the board they would pass a corner square marked "Labor upon Mother Earth produces wages" and collect a salary of one hundred dollars. Other squares on the board required players to pay tax, buy necessities, or take a Chance card. In one corner of the board was a square bearing the warning: "No trespassing. Go to jail." This space, she explained in an article for the *Single Tax Review,* was owned by a British lord and represented "foreign ownership of American soil." Anyone who landed there would be sent to the jail in the diagonally opposite corner of the board where they would stay until they rolled a double or paid a fifty-dollar fine. The final corner square contained a public park and the poor house where bankrupted players would be sent. Players could only leave the poor house if another player lent them enough money to clear their debts. After players had gone around the board a fixed number of times the game would end and the player with the most money would be declared the winner.

This is the game that would become Monopoly. But Magie's vision for the game was a world away from the internationally famous game it spawned.

The Landlord's Game, Magie explained, showed "how the landlord gets his money and keeps it." By playing it she believed children would learn "the quickest way to accumulate wealth and gain power is to get all the land they can in the best localities and hold on to it." She admitted that some might think this was a dangerous lesson but argued that the game would enable children to "see clearly the gross injustice of our present land system" and grow into adults who would campaign against it.

But just in case the message wasn't clear enough, Magie created an alternative set of rules designed to show how a single tax would create a more equal society. In this version of the game there was no poor house, no need to buy necessities, and the "no trespassing" space could be developed into a "free college" so that players would no longer be imprisoned. The way rent worked also changed. The rent due on undeveloped lots now went to the treasury instead of the property owners, and players could only earn money from others if they developed houses on their land. These rules, however, were optional. The core game remained one where players sought to get rich from owning land and despite the schadenfreude inherent in taking other players' money, Magie firmly believed that when people played the Landlord's Game they would see the injustice of it all.

After patenting her game in 1904, Magie began making copies by hand for other single taxers and gave one of these homemade sets to the residents of Arden, Delaware.

Founded in 1900, the town of Arden was one of several real-life experiments in Georgist economics being funded by the soap millionaire Joseph Fels. The businessman had already backed a similar experiment in Fairhope, Alabama, and eventually hoped to use the lessons learned from these communities to create a Jewish homeland based on George's single-tax theory.

In Arden residents could lease but never own land. The rent they paid would only reflect the value of the land they occupied and all the money paid to the town would be reinvested in the community. Arden's alternative approach to life attracted not just single taxers but other radicals and nonconformists too, including socialists like Upton Sinclair, the author of *The Jungle,* the 1906 exposé of the U.S. meat-packing industry.

The people of Arden enjoyed playing the Landlord's Game. It was, after all, a game based on the very ideals that had attracted them to this fledgling New Castle County community in the first place.

Scott Nearing, an economics professor at the University of Pennsylvania who moved to Arden in 1905, was one of the residents who played the game. Thinking it would be a great tool for teaching his students about the effects of rent gouging, Nearing made a copy by hand and began using it in his classes. He didn't know what the game's real name was so he called it the Anti-Landlord Game. His students decided Monopoly or Business would be a better name, but they liked the game all the same, and some of them made copies so that they could play it whenever they wanted.

While Nearing was introducing the Landlord's Game to economics students, Magie was struggling to get the game formally published. First she cofounded a small game company and released a version in 1904 but sold few copies. After abandoning that approach she offered it to Parker Brothers, the country's leading game publisher. Parker Brothers didn't want her game. It's too political and too complex, they told her.

It was a fair criticism. The politics were undeniable, indeed that was the whole point, and compared to other games on sale at the time it was complicated too. Most games of the era differed only by theme. Players might be racing airships, climbing the corporate ladder, or reenacting the conquests of Napoléon Bonaparte but the rules were always the same: Roll the dice, move your counter, and hope to reach the end first.

Compared with the brain-dead competition, the Landlord's Game was like George Orwell's *Animal Farm* in a world filled with endless versions of *The Very Hungry Caterpillar*. And while the Landlord's Game languished in limbo, the single-tax movement that inspired it shriveled away.

For many of the Progressive causes founded around the same time as the single taxers' movement, the 1910s were a decade of success. Those fighting for votes for women, prohibition, and trust-busting laws to break monopolies like Standard Oil made great strides. In contrast the single taxers spent the decade watching their movement fall off the political radar.

First came the death of Fels in 1914. Without his largesse, the growth of single-tax communities came to an abrupt halt. Then came the Red Scare. After the communists seized power in the 1917 Russian Revolution, the American public turned against socialism and other radical movements, seeing them as the enemy within. Fear about "Reds" secretly fostering revolution took hold, and in this climate of paranoia the single-tax movement became linked with socialism in the public mind.

This was a mistake. George was no socialist, just someone who believed his ideas could make capitalism work better, and there was little love between him and Karl Marx, the coauthor of *The Communist Manifesto*. Marx thought the introduction of a single tax would work against communism and in an 1881 letter he described George as "utterly backward."

George's take on Marx was no less rude. He believed communism would lead to dictatorship and called Marx "the prince of muddleheads." But in the grip of Red Scare paranoia, many failed to notice the gulf between George and Marx.

Yet even as the single-tax movement fizzled, the Landlord's Game was quietly winning converts within the halls of academia. Since Nearing introduced the game to his students, it had spread to universities across the northeastern states. Few of those who encountered the game knew where it came from or who made it or why, but they played it, liked it, made copies, tweaked the rules, and introduced others to it.

In an age of mass production this was rather quaint, a throwback to centuries past where games such as chess and backgammon spread via word of mouth and handcrafted copies while being refined slowly by the input of countless, nameless individuals.

People would make copies of the Landlord's Game for their friends on sheets of oilcloth that they carefully colored with paint or crayon.

They would type or write Chance cards and property deeds on unlined index cards and turn earrings, coins, and other miscellaneous household items into playing pieces. For the houses they would draw tiny homes on card stock and cut them out with scissors or use tiny pieces of painted wood.

As they re-created their own sets they also began tweaking the game and its rules. The properties on the board were often named after places near the home of the person who made the copy of the game. Properties were also arranged into groups so that players who owned all the lots in a set could charge double the rent. The "buy necessities" spaces were dropped, as was the limit on how many times players could go around the board. In a nod to the rise of the automobile, the public park became free parking. The single-tax version of the rules was forgotten and Mother Earth became Start or Go.

In 1927 the modified game ended up in the hands of Daniel Layman, a student at Williams College in Williamstown, Massachusetts. Layman introduced the game to his friends Ferdinand and Louis Thun. After playing it for a while, the Thuns came up with a new addition to the game: Community Chest cards. While the Chance cards mainly moved players around the board, the Community Chest cards usually gave money to players.

The brothers got the idea from real-life community chests, volunteer-run groups that would collect donations and then distribute the money to local good causes. The first community chest was founded in Cleveland, Ohio, in 1913, and the idea quickly spread across the country. By 1927 there were more than three hundred community chests and together they were distributing more than sixty million dollars every year. Corporate donors particularly liked the community chest model as it meant they could give without having to go through the trouble of identifying which causes to donate to.

After graduating in 1929 Layman returned home to Indianapolis and began playing the game with his friends there. Encouraged by the positive reaction, Layman persuaded a local battery manufacturer to publish his game, and in 1932 it went on sale as Finance.

The differences between Finance and present-day Monopoly were largely cosmetic. There were no hotels but players could build up to five houses on their land. Instead of the Just Visiting space next to the prison there was a tax demand for twenty dollars. Go was called Start. But beyond that and a few optional advanced rules, Finance was just Monopoly without the graphic flair.

The political dimension of the Landlord's Game was also missing. Finance's rules described the game as one that "parallels the transactions of modern business" and "gives everyone an opportunity to make a fortune."

Yet even with the politics gone, retailers were not convinced about Finance. Just as Parker Brothers told Magie twenty years earlier, stores regarded the game as too complicated and declined to stock it. Eventually Layman sold his rights to the game to an Indianapolis company called Knapp Electric who also found it a hard sell.

But Layman's game did find a fan in Ruth Hoskins. Before Finance's release, Hoskins had played a handmade copy of the game, then still being called Monopoly, while visiting family in Indianapolis. She thought it was great and made a copy to take back to Atlantic City, where she taught at a Quaker school. Hoskins and her Quaker chums then made a version of the game that reflected the resort city around them. The fifth house became a hotel. Layman's hotchpotch of real and imaginary streets became a tour of Atlantic City, from the run-down Baltic Avenue, where the African American maid of one of the Quakers lived, to the famous Boardwalk via Marven Gardens, a housing development a couple of miles to the south.

The railroads matched those that served the city too, including the Shore Fast Line, which connected the boardwalks of Atlantic City and Ocean City. The Shore Fast Line was one of the last interurbans, the electrified railways that bridged the gap between streetcars and proper railroads by using fast trolleylike cars to ferry passengers between different urban areas. During the early decades of the twentieth century interurbans thrived, only to be cut down in their prime by the arrival of affordable automobiles and the onset of the Great

Depression. By the time the Shore Fast Line was added to the board in 1932, most of the interurbans in the United States had shut down. In 1948 the Shore Fast Line would suffer the same fate.

The Atlantic City version of Monopoly spread to Philadelphia where it was introduced to Charles Todd and his wife, Olive, who lived in the city's Germantown neighborhood. Like so many before Todd enjoyed the game enough to make a copy for himself. The board he re-created on oilcloth was identical to the Atlantic City game except that he renamed the Shore Fast Line the Short Line and misspelled Marven Gardens as Marvin Gardens.

Shortly after making his copy, Todd bumped into Esther Jones, a childhood friend he hadn't seen in years. It was a chance meeting that would transform the game's fortunes.

After discovering that Jones lived nearby and was now married to a man named Charles Darrow, Todd invited the couple over for dinner. The dinner went well. Darrow, a brawny fortysomething who wore wire-framed spectacles, seemed a nice guy, direct but also personable.

They spent much of the evening discussing the economic woes of the city. The Great Depression had devastated Philadelphia. People had been reduced to begging or searching trashcans in the hope of finding enough leftovers to feed themselves. All over the city there were homeless encampments. People called these shantytowns Hoovervilles in honor of the president who presided over the Great Depression. Across the city around three hundred thousand people were out of work and Darrow was one of them. Since losing his radiator repairman job, Darrow had struggled to find work. He tried dog walking, repairing electric irons, even making and selling jigsaw puzzles, but these odd jobs brought in little money. Instead the family got by on the money his wife made from her needlework.

To make matters worse Dickie, the youngest of their two sons, had brain damage from a bout of scarlet fever. In 1930s America there was little support to help the family cope with the challenge of caring for their disabled son. The institutions that took in kids like Dickie were brutal places, places where people would often be chained to beds, beaten, poorly fed, and subjected to medical experiments. Unwilling to put their boy in such a place and unable to afford more humane care, the Darrows struggled on, trying to do the best they could for Dickie.

When the night was over, Todd and his wife told the couple that they must come again to play this game Monopoly. The Darrows had never heard of it but agreed to play it next time they visited.

When the couples next met, Todd got out his handmade Monopoly set and they played. A few days later Darrow asked Todd to make a set for him, and a little later, Darrow asked Todd to write up the rules for him.

Once Todd gave Darrow the rules they never spoke again. Todd was puzzled by the sudden loss of contact but then he spotted a poster announcing that Darrow would be demonstrating his new game Monopoly at a local bank. Todd was furious. Darrow had taken the game and was now selling it. How could he betray them like that? Todd fumed.

After getting the game and rules from Todd, Darrow had asked his cartoonist friend Franklin Alexander to bring some life to the dull board. Happy to help his fishing buddy, Alexander added colored bars to the property spaces and created a few illustrations that Darrow added to the board. The redesign made the game look a lot more exciting and attractive than the functional one Todd had made. Board enhanced, Darrow added the words "Copyright 1933 Chas B. Darrow" to the board and began making copies to sell. Nothing else had changed. Even Todd's misspelling of Marven Gardens as Marvin Gardens remained intact.

Darrow's early copies of Monopoly were made on a shoestring budget so tight that the game didn't even come with player tokens. Instead the rules suggested people use random household objects such as thimbles, coins, or trinkets from charm bracelets.

After selling a hundred or so handmade copies, Darrow used the money he had earned to pay for another five hundred sets to be manufactured professionally and then persuaded the Wanamaker's department store in downtown Philadelphia to stock the game. With Monopoly now being sold by the city's leading department store it became easier to get other retailers interested in the game, and soon after he got the toy store F.A.O. Schwarz to add it to their roster of board games too. Darrow also sent copies of Monopoly to the country's top game publishers, Parker Brothers and Milton Bradley, in the hope that they would buy the game from him. Both companies declined. Parker Brothers felt Monopoly was too complicated, took too long to play, and that concepts like mortgages would be alien to most players.

Darrow pushed on regardless. Not that he had much of a choice. He needed to make a living and Monopoly was the best chance of earning one that he had come across in years.

By October 1934, both Wanamaker's and F.A.O. Schwarz had sold out of the game and ordered more copies. Orders from other retailers were trickling in too, so Darrow paid for another 7,500 copies to be produced. Word of Monopoly's sales success in Philadelphia soon got back to Parker Brothers, which then offered to buy the game from Darrow after all.

On March 18, 1935, Darrow arrived at the company's office in the Flatiron Building in New York City to strike a deal with Parker Brothers president Robert Barton. Darrow would get royalties and seven thousand dollars on signing, they agreed. As they prepared to sign the contracts Barton asked Darrow if he was the sole inventor of the game. Darrow said yes.

After striking a deal Parker Brothers immediately set to work on fixing the game's most glaring deficiency: its lack of playing pieces. Building on the idea of using charm bracelet trinkets, Parker Brothers asked Dowst Manufacturing, a Chicago company that made toys for Cracker Jack boxes, to supply them with die-cast metal tokens for players to use. Soon every Monopoly set came with a battleship, a cannon, a flatiron, a shoe, a top hat, and a thimble. Over the next year or so Parker Brothers continued making improvements. It redesigned the paper money, added illustrations to the Chance and Community Chest cards, and created a mascot for the game: a stereotypical financier who looked like a cuddly J. P. Morgan and is known today as Mr. Monopoly.

As soon as Parker Brothers took over production of Monopoly the company found itself besieged with orders. Word about the game was spreading fast. Monopoly was hot stuff but no one really knew why. Was it the thrill of being able to hold fistfuls of paper money after years of struggling to scrape together pennies? Or maybe it was the ability to buy and own property at a time when most Americans rented their homes. Maybe the cutthroat nature of the game gave people a release. In Monopoly they could hammer their friends and relatives into bankruptcy without having to worry about repercussions because it was all part of the game. Honest.

Or could it be that Monopoly finally gave adults a board game they could relate to?

"Even though we see Monopoly today as a family game it was primarily played by adults," says Philip E. Orbanes, who became Parker Brothers' resident expert on the game during his time as a senior vice-president of the company. "By today's standards Monopoly might be heavily luck orientated but it was revolutionary at the time. As a result of its success, Parker Brothers, Milton Bradley, and all their major competitors had the confidence to try other adult-orientated games."

With sales soaring, Parker Brothers decided to patent the game but the company's lawyers soon found a problem. It wasn't Darrow's game. While preparing the patent application the legal team found Magie's patent on the Landlord's Game and Layman's very similar Finance game. They also found another version of the game called Inflation on sale in Texas. Parker Brothers responded by getting out its checkbook. The company paid ten thousand dollars to take control of Finance and also bought the rights to Inflation. But, most importantly of all, the company also bought the rights to the Landlord's Game.

By this point Magie was living in Arlington, Virginia, and the deal was so important to the future of Monopoly that George Parker, the company's sixty-nine-year-old founder, went to strike a deal with her in person. Magie was delighted to see him. Finally, after all this time, someone wanted her game. She agreed to sell the rights to Parker Brothers in exchange for five hundred dollars and a commitment that the Landlord's Game would be published too.

The company also used the situation to renegotiate its deal with Darrow and reduce his royalties on the game. But to the outside world nothing had changed. Parker Brothers maintained the pretense that Darrow was the creator of Monopoly. His rags-to-riches story was media catnip and far easier for the public to relate to than a tale about a eccentric woman inspired by a long-forgotten economist and countless iterations by people unknown.

Even with his reduced royalties Monopoly made Darrow a millionaire. By the end of 1935, more than 250,000 copies of the game had been sold. A few months later Darrow retired and the family moved to a farm in Bucks County, Pennsylvania. He spent the rest of his life traveling the world, growing orchids, and cultivating thickets of roses to protect pheasants from hunters. And while his wealth could never restore Dickie's health it did allow the Darrows to give him the very best care money could buy.

Even greater success was to come. "On January 2, 1936, the deluge came down upon Parker Brothers in the form of such an overwhelming demand for the game that our modest factory was immediately put on a three-shift twenty-four-hour basis," Barton recalled in a 1957 letter. "We ceased the publication of almost every other game in our line."

It wasn't enough. Parker Brothers were soon producing twenty thousand Monopoly sets a week and merely treading water. Orders from retailers arrived at such a rate that the halls of the company's headquarters in Salem, Massachusetts, became cluttered with wicker laundry baskets filled to the brim with outstanding orders. Overwhelmed, the company asked external accountants for help. One firm they approached took one look at the hallways crowded with order-filled laundry baskets and turned the job down on the spot. Not even the decision to jack up the price of Monopoly by 25 percent could curb demand.

That year Parker Brothers sold 1,750,000 Monopoly sets in the

United States. Monopoly mania would ease but the game would keep selling hundreds of thousands of copies a year until the Second World War forced production to be curtailed.

In 1939 Parker Brothers finally got around to putting the Landlord's Game into production. Magie was excited. She liked the new look Parker Brothers had given her game and hoped that at long, long last the Landlord's Game and its message would finally find an audience.

Given her high hopes, what followed must have been heartbreaking. Sales were dire. Retailers who stocked the Landlord's Game began threatening to stop carrying Monopoly unless Parker Brothers took the game back. Most of the ten thousand copies of the Landlord's Game the company produced were destroyed. Magie could only look on helpless as her board game dream fell apart. The Landlord's Game was dead. Her bid to spread the gospel of Henry George had failed. All that was left was Monopoly, the Frankenstein's monster she had inadvertently created.

The message of the Landlord's Game still lurks in Monopoly. It can still be seen in the way that every game ends with one rich monopolistic landlord and everyone else ruined. But people did not, as Magie hoped, see the injustice.

Instead, players looked at Monopoly and decided they wanted to be the rich monopolistic landlord. After all, who wants to be bankrupt? Much better to be the one *doing* the bankrupting. If winning the game meant bleeding your opponents dry so be it. If Monopoly seemed like a celebration of dog-eat-dog capitalism, that's because that is really what people wanted it to be.

In the years that followed, Monopoly's embrace of ruthless competition would prompt bans in communist China and the USSR. After taking over Cuba, Fidel Castro called the game "symbolic of an imperialistic and capitalistic system" and ordered every copy on the island to be seized and destroyed. Such was the communist opposition to the game that in eastern Europe illicit homemade copies of Monopoly circulated in secret among those who dreamed of freedom and capitalism rather than dictatorship.

None of that stopped Monopoly. By 2016 it had sold more than 250 million copies worldwide. It is, by far, the bestselling branded board game ever created and no other game, except maybe chess, has so imprinted itself on the world's collective consciousness. The game's terminology has seeped into our language. Phrases like "Get out of jail free card," "Do not pass go," and "Collect $200" are familiar to almost everyone.

Who knows what else Monopoly has massaged into our minds through all those billions and billions of games? After the Second World War, home ownership boomed and the United States went from being a nation of renters to a land of homeowners. Much of this was due to economic policy and rising incomes, but it's not unthinkable that Monopoly planted the thought in people's minds that owning property is the path to wealth, and that no one wins by paying rent.

Assuming we played it correctly that is. For despite its success, a surprising number of Monopoly players don't know the rules. "There's a lot of people who have always played Monopoly wrong," says Ben Rathbone, vice-president of design at Monopoly's present-day owners, Hasbro Gaming, which acquired Parker Brothers in 1991. "There are people who are convinced that any fines or taxes go into the center of the board and if you land on Free Parking you get it. It's so bad that we had a copywriter once, years and years ago, who wrote it into the rules. We were like 'No, no, no,' and she was arguing that it was in the rules and we were like, 'No, it's not.'"

The trouble with the Free Parking rule is that it makes the game last forever. "People think it makes sense because they get more money but what they don't realize is that Monopoly is designed so that you've got to get money out of the game," says Rathbone. "If you start putting money back in it makes it longer, because you have to bleed people dry."

The time it takes to play Monopoly has dogged the game for years, says Orbanes: "While the game found favor, one of Parker Brothers' original criticisms of the game still stands: that it takes too long to play."

This is a bigger problem today than it was in 1935. In the mid-oughties research by Hasbro's marketing department found that, compared to previous decades, families no longer spent enough time together to learn and play new games.

"The family as a nuclear unit was together for far less time than at any period in history, and so what little time people did have together was precious and they wanted a complete experience," says Orbanes. "We began to define these periods of togetherness as 'pushed.' We needed to create things that could be completed in a burst. A burst might mean thirty or five minutes but whatever it meant, we had to accommodate the reality of it."

For Monopoly this was a big problem. "Hasbro's Dave Wilson, who was the president at the time, actually came to me lamenting this because he didn't know the way out of the box: Monopoly took as much time as it did because it was Monopoly," says Orbanes.

Then one morning in the shower Orbanes came up with the answer: the speed die.

After players had gone around the board once, they would start rolling the speed die as well as the two normal dice. If the speed die showed a one, two, or three, that would be added to the number of spaces the player could move. If the speed die showed a bus, players could use the total or one of the individual numbers shown on the two normal dice when moving. Finally, if the speed die showed the game's mascot Mr. Monopoly, players would be whisked to the nearest property still for sale, and if everything had been bought they would be sent to the next space where they would have to pay rent.

Orbanes didn't have regular Monopoly in mind when he created the speed die. He was actually thinking that it would be good for Monopoly: The Mega Edition, an enlarged version of the game he was working on that was released in 2006, but Hasbro liked the way the speed die accelerated the game and made it a permanent feature of standard Monopoly in 2007.

Six years later Hasbro went one step further with Monopoly Empire, a Monopoly spin-off so speedy that games can flash by in less than thirty minutes. Instead of buying land and building houses, players now purchase big-name brands like Coca-Cola, Spotify, and My Little Pony in a race to fill up their advertising billboard tower that takes the viciousness of Monopoly to new heights. The game is a frenzy of backstabbing and one-upmanship. Players steal brands from one another's towers, sabotage purchases, and get carted off to jail for insider trading. Opponents trip each other up with "rent reversal" cards and form short-lived alliances to bring down their rivals.

"We wanted to do something very different," says Rathbone. "It's essentially the same game but faster, more cutthroat and people loved it because it felt like Monopoly but had different game play. So we are playing with more things like that."

With its blatant corporate branding and savage heart, it is hard to imagine a game more at odds with the ideals of the protest game Magie created back in 1902 than Monopoly Empire. So it's probably just as well that Magie is no longer around to see what her anti-landlord parable became.

6

# FROM KRIEGSSPIEL TO RISK: BLOOD-SOAKED AND WORLD-SHAPING PLAY

## *How board games prepared the world for war*

The first limousines arrived at the imposing iron gates of the Naval War College in Tokyo just before nine in the morning. More and more vehicles followed, unloading admirals, generals, and senior military officers.

Inside the college, the growing crowd of Japan's military elite buzzed with excitement. They always enjoyed these weeklong annual gatherings. In the daytime they would play tabletop war games designed to put their offensive and defensive strategies to the test. Then they would dine and drink well into the night.

For most of those attending the September 1941 Tokyo war games the event seemed to be the same as always, but in a quiet corner of the college's east wing something was different: a room strictly off-limits to anyone without an invitation.

Inside were thirty or so of imperial Japan's most senior officers, and they had gathered to play a top-secret war game that would change the course of history: a rehearsal of a wild plan to launch a surprise attack on Pearl Harbor.

In the middle of the room stood a long table covered in papers. On the walls hung maps of the Pacific Ocean on which the Hawaiian island of Oahu and its naval base, Pearl Harbor, had been highlighted.

All around the room were books and folders filled with data that had been gathered by Japanese spies, together with complicated tables of probability that would govern the outcomes of the game they were about to play.

The officers in the room were divided on the Pearl Harbor plan. Vice Admiral Isoroku Yamamoto and his supporters believed the strike could neutralize the threat the United States posed to Japan's ambition of taking control of East Asia. But most of the officers, including Vice Admiral Chūichi Nagumo, thought the plan was too crazy to work. Even many of Yamamoto's own deputies thought the idea was nuts, although they said nothing out of loyalty to their superior officer. The attack would almost certainly fail, the opponents reasoned, and trying it would divert vital naval forces away from the invasion of East Asia.

But the time for debate was over. This tabletop war game would be the nearest they could get to testing the Pearl Harbor mission without actually doing it for real.

After an introductory talk on how the game would work, the officers divided into two teams. One would command the Japanese fleet, the other the defending Americans. The teams relocated to separate rooms so that they would remain unaware of what the other was doing and began playing, issuing their orders to the umpires in the main room who had the task of overseeing the game and relaying information back to the teams.

The game ended in disaster for Japan. Team Japan tried to sneak their fleet across the Pacific, only for a U.S. air patrol to notice oil on the surface of the water that had leaked from a submerged Japanese submarine. After seeing this the U.S. team got twitchy and expanded the range of their aerial patrols from three hundred to six hundred miles. Soon after, a U.S. reconnaissance plane found the naval task force heading toward Oahu. The Japanese shot it down before it could report its discovery but the element of surprise had been lost. Now it was only a matter of time before the Americans figured out where they were.

The Japanese fleet accelerated, racing to Oahu as fast as they could. As the fleet dashed across the seas word came in from their submarines: ten American cruisers were now heading in their direction. About two hundred miles from Oahu, the Japanese fleet pulled to a halt and launched its first attack wave. It failed. American antiaircraft guns, ship cannons, and interceptors blasted the Japanese fighters and bombers out of the skies.

The Japanese tried again, launching a second attack that proved no more effective than the first. The few surviving Japanese planes returned to the fleet having inflicted only minor damage to the Oahu naval base. The Japanese ships began to flee but American bombers soon caught up with them and sank a third of the armada.

The attack on Pearl Harbor had backfired. Japan was now at war with the United States and its naval power had been severely weakened in the process. Just as we expected, the operation's opponents thought. But this was just the first game and when the teams regrouped for a second go the following day, the Japanese team had learned the lessons of its failure.

This time the Japanese players changed their route, sailing to a point 450 miles north of Oahu. On reaching it the ships began heading due south at top speed toward the island as the sun began to set. Aware that American air patrols usually returned to base around sunset, the Japanese team figured that they would get within striking distance of Pearl Harbor before the next patrol could spot them.

The gamble worked. The fleet arrived two hundred miles to the north of Pearl Harbor undetected and launched a devastating surprise attack that sank four battleships, two carriers, and three cruisers. A battleship and four cruisers were left seriously damaged. American air power also took a beating. One hundred U.S. fighters were shot down in the air and eighty more were destroyed before they could even leave the tarmac.

Damage done, the Japanese fleet then escaped from the scene with help from a timely squall with only one carrier lost and some minor damage. The second game had proved that attacking Pearl Harbor was not so crazy after all.

A month later the officers returned to the war college and played again with a smaller Japanese armada, a new rendezvous point for the ships, and questions about the refueling of the vessels resolved. Once more the Japanese team launched a successful surprise attack and emerged victorious, strengthening the case for attacking Pearl Harbor yet again.

On December 7, 1941, the Japanese put the lessons from their war games into action. The fleet sailed the northerly route devised at the war college and delivered a powerful strike that sank eighteen ships, destroyed nearly two hundred planes, and killed more than two thousand Americans. The Japanese retreated from Oahu with comparatively minor losses.

The attack rehearsed in secret on the war college tabletops changed the course of history, drawing the United States into the Second World War and paving the way for the Third Reich's defeat in Europe and the development of the atomic bomb.

Japan's use of a game to plan such an attack was anything but unusual. At the time all the world powers used similarly complex board games to put their strategies to the test.

The idea of using games as military planning tools originated in the Germanic states of Europe. One of the earliest attempts came in 1559 when Count Reinhard zu Solms, a military theorist from the Hessian town of Lich, devised a card game to help with the planning of military formations. Each card represented different military units and players would arrange them into competing formations before debating which was the best. As a military tool and as a game it was a disappointment. There were no rules and no way to clearly determine who had the superior arrangement of troops. Nevertheless, it was start.

Undeterred, the German fascination with war games continued for the next two centuries. Elaborate versions of chess played on enormous boards with hundreds, even thousands, of squares and new types of playing pieces abounded. None, however, proved useful to the military. The pieces still moved around the board in fixed, unnatural ways and captured each other as in chess regardless of what military unit they represented.

By the end of the eighteenth century the need for a better war game was intensifying as firearms increased the need for strategies that went beyond having masses of troops slugging it out in hand-to-hand combat.

Slowly but surely war games edged away from chess and toward realism. In 1780 the German mathematician Johann Hellwig introduced squares that represented different types of terrain. Sixteen years later Giacomo Opiz of Bohemia replaced chess-style combat and began using dice to determine the outcome of battles.

The big breakthrough came just four years later in 1810 when Lieutenant Georg von Reisswitz of Prussia turned his hand to creating a game of military merit. While others offered flat boards neatly divided into equal-size spaces, he built a three-dimensional model landscape of hills, rivers, and forests to play on. No longer would players face tabletop battlefields where the rivers bent at right angles and hills were square.

Crucially, he also made the board to scale so that three centimeters on his board represented one hundred paces. This being an age where armies still moved on foot, the pace was the most important military measure around—a way to calculate how fast and far troops could travel. A single pace equated to two and a half feet and when marching in quick time troops were deemed capable of moving at a rate of 120 paces a minute.

The use of scale and the freeform board would have been major innovations on their own, but von Reisswitz didn't stop there. He also decorated dozens of small wooden cubes with symbols so that they represented different military units and created mathematical rules to govern their movement.

The rest of von Reisswitz's rules, however, were so light of touch they covered almost nothing other than movement. Instead everything else in the game would be decided by an umpire who, like a precursor to a Dungeons & Dragons dungeon master, would determine the outcome of every clash and the effects of the minutiae of warfare, such as how long it would take for new commands to reach the troops.

Von Reisswitz didn't have any grand plans for his game. He, rather unimaginatively, called it Kriegsspiel, the German word for "war game," played it with some friends and left it at that. But then a captain who knew of his game mentioned it during a lecture to army cadets at the Berlin Military Academy. In the audience were the sons of King Friedrich Wilhelm III. Intrigued the two princes invited a surprised von Reisswitz to Berlin Palace to show them his creation.

Von Reisswitz arrived at the vast baroque palace and was escorted to the grand White Hall, a large chamber decorated with white marble tiles and silvered ornaments. There waiting for him were the two princes and a group of royal advisers.

After von Reisswitz's demonstration of the game, the excited princes wrote to their father about their find. Shortly after von Reisswitz was invited to demonstrate Kriegsspiel to the king in Potsdam, home to the royal residencies of the Kingdom of Prussia. Fearing his poorly constructed game wouldn't survive the journey to Potsdam, von Reisswitz said he would make a superior version and bring it to the king as soon as possible.

A year later, von Reisswitz finally arrived at Sanssouci Palace, the grand rococo-style summer residence of the king, with a large chest that measured six feet square. On the top of the chest was a large wooden board and beneath it drawers filled with playing pieces and the equipment needed to play Kriegsspiel.

Among the items in the drawers was a collection of terrain pieces made of plaster and painted to represent different kinds of terrain, from roads and villages to swamps and hills. These four-inch-square tiles were designed so that they could be endlessly arranged and rearranged to form new landscapes for players to battle on.

The wooden cube troops, meanwhile, were now porcelain figurines. In other drawers were the tools needed to play the game, including a ruler for measuring distances and tiny boxes for hiding troops that the opposing player's forces shouldn't be able to see.

Impressed, the king ordered the game to be permanently set up in the palace and over the next few years it became one of his favorite pastimes. He began holding Kriegsspiel parties and his doubtless long-suffering family and courtiers often had to stay up well past midnight playing it with him. The king never used the game to plan specific battles but he did credit it with giving him ideas for army maneuvers. But as the years passed, the king spent more and more time away from Potsdam and the game sat un-played.

While the king might have forgotten about the game, von Reisswitz's son Georg had not. Like his father, the younger von Reisswitz was a military man. He joined the artillery corps at the age of fifteen, around the same time as his dad was showing the game to the Prussian princes, and soon after he founded a Kriegsspiel club at the Berlin garrison where he and other officers would play his father's game once or twice a week.

Over the years the younger von Reisswitz made improvements to the game that he tested out on those who attended his Kriegsspiel club. He replaced the tiles with military maps of real terrain so that the tabletop battles had direct relevance to actual conflicts. He greatly expanded the rules too, so that they governed matters as diverse as nighttime attacks and the construction and destruction of bridges. He also used military data about weapon performance to create tables of probability that players could then, with the help of dice, use to calculate the effects of artillery and gunfire on enemy troops.

Word of the younger von Reisswitz's much improved Kriegsspiel soon reached the royal court and he too was invited to Berlin Palace to demonstrate his version of the game in early 1824. Shortly after that meeting he was ordered to report to Karl von Müffling, the chief of the Prussian General Staff.

On arriving von Reisswitz discovered that the entire top tier of the Prussian military had come to see his creation. At first von Müffling seemed disinterested in the young artillery officer's game but as the demonstration progressed he became absorbed, eventually declaring: "This is not a game, this is a war exercise. I must recommend it to the whole army!"

A few days later, von Reisswitz was put in charge of a workshop tasked with assembling Kriegsspiel sets for every regiment in the Prussian army. By fall 1824 Kriegsspiel had become part of Prussian military training and clubs dedicated to the game existed in regiments across the kingdom.

Despite the enthusiasm of the top brass for Kriegsspiel, not everyone within the army welcomed the game. Older generals, especially, felt put out by Kriegsspiel, feeling that it devalued the experience they had gained on real battlefields. They also hated the way it made younger officers think they knew what they were doing. The only wars they knew involved pushing toys around tables. They knew nothing of blood and horror and the smoke of artillery fire.

The old guard also disliked von Reisswitz. They found his confidence and jokey manner irritating and bristled at the acclaim being heaped on this low-ranking officer. So they put him in his place. They promoted von Reisswitz to captain and transferred him to an artillery brigade some 150 miles south of Berlin in Torgau. It was banishment and von Reisswitz, who wanted to stay in Berlin, took the move badly. His cheerfulness deserted him and in the summer of 1827, while on home leave, he ended it all by taking his gun and shooting himself.

While von Reisswitz did not live to see it, the generals who opposed Kriegsspiel lost the argument, and his game became one of the key innovations that helped turn Prussia into one of the world's major military powers as the 1800s wore on.

By the 1850s Kriegsspiel had become a central tool in Prussian military planning and officer training. And when the Franco-Prussian War broke out in July 1870 the benefits of those countless hours spent at the Kriegsspiel table were demonstrated to the entire world.

The war was sparked by France's growing concern about Prussia's desire to create a united Germany. Fearing the creation of a new central European empire on its doorstep, France declared war. The

French expected to prevail, not least because it had superior weapons, including the fearsome mitrailleuse, an early form of machine gun, and most of the world believed the same.

So it came as something of shock when the exact opposite happened. In battle after battle the Prussian military delivered fast and decisive victories. By September the Prussians had swept through northern France and surrounded Paris. In January 1871 Paris surrendered and the Germanic states united as Germany.

In the aftermath, the Prussians boasted about the role that Kriegsspiel played in their stunning victory, crediting the game with enabling their officers to act fast because they had already practiced countless battles on the tabletop. Other militaries took note and began adopting Kriegsspiel. Soon the Austrians, the Russians, the British, the Italians, the Americans, the Japanese, and yes, even the French were introducing their commanders to the German war game.

And as the world headed toward the seemingly inevitable Great War, Kriegsspiel was deployed to refine the strategies that would be put into action should this "war to end all wars" erupt.

In the early 1900s the U.S. Navy used Kriegsspiel to simulate a clash with British naval forces. To their horror they found the longer ranges of the guns on British ships meant U.S. naval vessels could be attacked from distances where they could not even hope to fight back. As a result the U.S. Navy began developing longer-range guns and reinforcing its ships to give them extra protection in case they went to war with the British.

Not every country took the lessons from Kriegsspiel so seriously. After Japan and Britain formed an alliance in early 1902 in response to Russian expansion in northeast Asia, the Russians organized a war game to assess the outcome of war with Japan. The game suggested that rather than declaring war the Japanese would launch a surprise attack that would destroy the Russian fleet at Port Arthur, the northerly port now known as Lüshun that Russia had leased from the Chinese.

This finding should have set alarm bells ringing in the Russian military. In winter the seas around most Russian ports turn to ice, trapping their naval vessels, so losing access to the warmer waters of Port Arthur would be a major blow. But the warning from the game went unheeded. The Russian military was skeptical about Kriegsspiel, dismissing games as being for little children and feeling that all those dice rolls and probability equations weren't worth the effort.

The Japanese took war games more seriously and their games had come to nearly the same conclusion as those played by the Russians. In February 1904 Japan launched a surprise attack that crippled the Russia's Pacific fleet. The Japanese won the resulting Russo-Japanese War and forced Russia to leave Port Arthur, Korea, and Manchuria. After that the Russians took Kriegsspiel more seriously.

While everyone else was catching up with Kriegsspiel, Germany was busy reinventing the game. Following its 1871 victory over France the German military broke away from the rule-heavy games devised by the younger von Reisswitz in favor of "free Kriegsspiel." The new incarnation aimed to make the most of the firsthand battle experience German generals had gained during the Franco-Prussian War. It slimmed down the tedious calculations and upped the importance of professional opinion in determining how the battles on the board worked out. In many ways it was a return to the original Kriegsspiel developed by the elder von Reisswitz in 1810.

And it was this version of the game that Field Marshal Alfred von Schlieffen, the chief of the German General Staff, began using in 1897 to develop a way for Germany to win a war against France and Russia. Such a scenario was a major concern for the Germans. Germany was sandwiched between these two great empires and as such vulnerable to a simultaneous invasion from France and Russia.

Von Schlieffen played hundreds of games of Kriegsspiel that replicated that very scenario and the lessons he learned informed a strategy that became known as the Schlieffen Plan.

The plan rested on neutralizing France as fast as possible so that the German army could then focus on dealing with the Russians on its eastern front. In order to defeat France quickly, the Schlieffen Plan proposed sending German armies through Belgium and the Franco-German border so that the French forces could be encircled and then defeated. The games von Schlieffen played suggested his plan would deliver victory over the French in just six weeks.

Although von Schlieffen retired in 1906, his successor General Helmuth von Moltke continued to refine the plan with the help of war games. In one session von Moltke noticed that Germany's armies would run out of bullets and shells before the French were defeated so he ordered the creation of the world's first motorized ammunition battalions to keep the guns blazing.

But for all their planning, when Germany put the Schlieffen Plan into action at the start of the First World War, the real world refused to conform to the tabletop version of events. The Russians mobilized quicker than their games predicted, forcing von Moltke to divert troops east, and the Belgians resisted the invasion more forcefully than expected too, slowing the Germans' march to France. Then, as the German forces pushed into France, a unexpected breakdown in communications left von Moltke in the dark about where his units were and unable to coordinate them effectively.

On the tabletop the Schlieffen Plan promised victory in six weeks. In reality it delivered a stalemate and four horrific years of trench warfare.

After the failure of the Schlieffen Plan it would be tempting to think Germany's love affair with war games was over, but in fact the opposite happened. Following their victory in the First World War, the Allies imposed strict limits on German troop exercises and so war games became more important than ever for the country's military. And after the Nazis took power they prepared for their conquest of Europe using war games to plan out everything from the invasion of Poland and Russia to the London Blitz, often with uncanny accuracy.

But after the Second World War the days of generals pushing panzers across maps spread over vast tables and consulting probability charts began drawing to a close. Six months after Japan surrendered, the University of Pennsylvania switched on the ENIAC, one of the world's first programmable computers. This thirty-ton colossus of a machine had one purpose: to calculate artillery-firing tables for the U.S. Army. It was a sign of things to come.

As processing power grew, more and more of the mathematical work involved in playing war games began being handled by computers. By the late 1970s computers that could do all the necessary calculations in a fraction of the time had made tabletop war games largely redundant.

Yet even as the armed forces swapped tables for screens, the war game found a new lease of life in our homes. The idea of playing war games for fun dated back to the early twentieth century, when people inspired by the military mania for Kriegsspiel began writing rules for war games that people could play at home with the aid of toy soldiers.

The most famous of these rulebooks was 1913's *Little Wars*. Written by H. G. Wells, the author of *The War of the Worlds* and *The Invisible Man, Little Wars* sought to rescue war games from the tedium of military number crunching.

As one might expect from a game created by a man who found the world's passion for chess "unaccountable," *Little Wars* prized action over strategy. Instead of painstaking calculations and realism, Wells advocated the creation of imaginative battlefields where cardboard forts sat next to rivers drawn with chalk and trees made from the twigs of shrubs. In place of dice and probability tables, *Little Wars* used simple rules of thumb to govern combat and encouraged the use of spring-powered toy cannons to inject the game with more action.

As a committed pacifist, Wells saw his game as an antidote to real warfare. "How much better is this amiable miniature than the Real Thing!" he exclaimed in his book. "Here is the premeditation, the thrill, the strain of accumulating victory or disaster—and no smashed nor sanguinary bodies, no shattered fine buildings nor devastated country sides, no petty cruelties, none of that awful universal boredom and embitterment, that tiresome delay or stoppage or embarrassment of every gracious, bold, sweet, and charming thing, that we who are old enough to remember a real modern war know to be the reality of belligerence."

Despite Wells's efforts to get the world to play, rather than wage, war, it would be another forty years before war games truly became a pastime.

The turning point came in 1952 when military history fan Charles Roberts joined the U.S. National Guard and began looking for war games to play in the hope that they would develop his understanding of warfare. Since war games were still rarely seen outside the military, Roberts was unable to find such a game. So he did the next best thing: he made a war game of his own. He named it Tactics.

Tactics simulated conflict between two fictional nations using tanks, infantry, and air power. Not unlike the chess-inspired games that predated Kriegsspiel, the board was a grid of squares, each of which depicted a different type of terrain from water to mountain range. Its combat, however, owed more to the work of the younger von Reisswitz. To learn how each bout of combat fared, players would roll a die and then consult a table that used the number rolled and the ratio of attacking to defending units for the result.

In 1954 Roberts decided to publish Tactics "for a lark" and began selling it via mail order from the garage of his home in Avalon, Maryland. Unlike many of the earlier war games for the home, which amounted to little more than a set of rules and the occasional map, Tactics came with all the miniature tanks, infantry, and planes needed to play it in its box.

Over the next four years Roberts sold around two thousand copies of Tactics and became convinced that there was an audience out there for games like his that was being overlooked by the industry. So in 1958 he founded Avalon Hill, a game publisher dedicated to producing complex games for adults. Roberts envisaged his company selling sports and business simulations as well as war games but when it came to sales it was clear the money was in military games. Over the next few years Avalon Hill would turn war gaming into a thriving, if niche, hobby.

In spring 1964 the company launched a self-promoting magazine called *The General,* which became the focal point for everyone interested in war gaming. *The General* offered free advertising to war-gaming clubs and in doing so helped to foster a community of armchair generals, who in turn began publishing fanzines and designing new war games. And as the hobby grew so did Avalon Hill's sales.

But while Avalon Hill successfully spawned the active but small war-gaming community, it would be a French film director who brought the genre to the masses.

In the early 1950s Albert Lamorisse, who would later win an Academy Award for his 1956 film *The Red Balloon,* devised a war game called La Conquête du Monde, or the Conquest of the World, during a family vacation. As the name suggests, the goal was world domination, and it had players waging war with Napoleonic armies and navies across the globe until only one remained.

Unlike Avalon Hill's more mathematically minded efforts, Lamorisse's game distilled the battles for each slice of territory down to a contest of dice rolls. With each roll of the dice, the player with the lowest score lost a unit and this continued until one side was eliminated.

After applying for a patent in March 1954, Lamorisse sold La Conquête du Monde to the French game publisher Miro, which asked game designer and philosophy teacher Jean-René Vernes to improve it. Vernes removed the naval vessels and reformed the combat so that it favored defending players. Now players would get one die for each army they sent into battle up to a maximum of three, and if they rolled the same number as the defending player they would lose one of their attacking armies.

Revisions made, La Conquête du Monde went on sale in France in 1957. Soon after, Miro showed the game to the U.S. game giant Parker Brothers, who jumped at the chance to release it in North America.

Despite snapping up the rights, Parker Brothers felt Verne's improvements did not go far enough. The main problem was that the game took ages to play as troops chipped away at entrenched defenders over and over again. To fix the problem Parker Brothers tipped the balance of combat in favor of the attackers by limiting defenders to using no more than two dice in combat. Problem solved, they renamed the game Risk and launched it in 1959.

Many within Parker Brothers were uneasy about Risk. Sales of war toys were in decline as parents turned against the idea of arming children with plastic guns and the other pretend paraphernalia of warfare. The company feared that Risk would also be a victim of this parental backlash. Price was also a concern. Most board games cost around two dollars but Risk cost seven and a half, thanks to its large board and the heaps of colored wooden pieces that formed the in-game troops.

Yet instead of scaring off customers with its price and military theme, Risk became one of the hottest games of the year with more than 100,000 copies sold in its launch year. In the years that followed Risk sold millions, becoming not only the most popular war game ever devised, but one of the most popular board games of all time.

Committed war gamers with lifetime subscriptions to *The General* might have sneered at the simplistic combat and lack of realism but their preferred games couldn't match the immediacy of Risk. "From the very first move it's competitive," says Philip E. Orbanes, who founded war-game publisher Gamescience in 1965. "You're butting up against somebody and you're attacking them or they are attacking you. It's not like you sit back and patiently build your army—it's instant action from the beginning and that's one of its great merits."

And thanks to its mass appeal, Risk continued to thrive even as tabletop war gaming slipped into seemingly terminal decline in the late 1970s.

The first blow to war gaming came from a monster of its own making: Dungeons & Dragons. First published in 1974, Dungeons & Dragons was created by war-game designers Dave Arenson and Gary Gygax. A few years earlier Gygax had codeveloped a medieval-era war game called Chainmail that included fantasy elements such as magic. Chainmail went down well with war gamers, but it was clear that it was the fantasy and not the medieval warfare that was winning them over.

Gygax and Arenson responded by making a pure fantasy game that transferred the dice and probability combat systems of war games into a setting inspired by J.R.R. Tolkien's *The Lord of the Rings*. The result was Dungeons & Dragons, the first role-playing game.

In Dungeons & Dragons players took on the persona of wizards, warriors, and other heroes and went on epic quests set up and managed by a dungeon master, who acted much like the umpires of Kriegsspiel games. But unlike the war games that inspired it, Dungeons & Dragons did not require a board. Instead it relied on players' imaginations. Aided by the storytelling of the dungeon master, players would act out their adventurers' lives and use dice to determine the outcomes of decisions and battles with fearsome monsters.

Dungeons & Dragons became a phenomenon and spawned countless other role-playing games from the cyberpunk adventures of Shadowrun to the gothic horror of Vampire: The Masquerade. By 1979 more than a quarter million Dungeons & Dragons rulebooks had been sold, and many of those who bought them might once have found themselves playing general in a war game rather than role-playing an orc-slaying elf.

The late 1970s also introduced another challenge to war games: the home computer. As the 1980s progressed computerized war games began to usurp the tabletop games that inspired them, just as they did in the military war rooms. The appeal of computerized war games was twofold, says Joel Billings, founder of Strategic Simulations, Inc., the California computer game publisher that was at the forefront of war gaming's digital transition during the 1980s.

"One was to have someone to play the game against because it was always hard to find opponents," he says. "I had spent most of my childhood playing solitaire. Occasionally I played by mail or at a store where you could connect with other gamers and play, but it was mostly a lot of playing solitaire. So being able to play against the computer, that was great. That was number one. And a close second was 'fog of war.'"

On a real battlefield armies do not have perfect knowledge of enemy positions and actions. That gap in awareness is called the "fog of war" and it had always been the Achilles' heel of tabletop war games. When playing a board game, maintaining the fog of war is near impossible. Even if your troops were covered with tiny boxes as in the original Kriegsspiel, your enemy still knew you had something there even if they did not know exactly what it was. The only effective solution to the problem is to keep players in different rooms and have an umpire managing the communication just as the Japanese did when planning the attack on Pearl Harbor. But playing against people who you can't even be in the same room with isn't much fun.

Computers, however, could hide information with ease by only displaying what players should know—replicating the fog of war experience.

In almost every other way the early computer war games were just like their tabletop counterparts with turn-based play and combat governed by probability tables. But the appeal of playing against the computer and fog of war were enough to draw many away from the tabletop games. By the dawn of the 1990s, tabletop war gaming was in steep decline, hemorrhaging players to computer war games and role-playing games. But while the hobby became even more niche, Risk continued giving people across the world a lightweight taste of war gaming both in its original form and through a multitude of spin-offs, some of which bare almost no relation to Lamorisse's original.

Risk: Star Wars Edition is a perfect example. Released in 2015 it bears the name of the world's favorite war game but it's hard to spot any other connection with the original Risk's Napoleonic warfare. Instead it lets two players re-create the climactic battle of *Return of the Jedi* by directing spaceships around the galaxy on a board shaped like Darth Vader's TIE fighter.

Yet while this tabletop space battle set in a galaxy far, far away and the abundance of computer war games on sale today might seem far removed from the war games created in nineteenth-century Prussia, their roots still belong to that blood-soaked, world-shaping board game genre.

# 7

# I SPY

## *How Chess and Monopoly became tools of espionage and propaganda*

"I would like to have some visiting cards printed, please," said the stocky man with the heavy eyebrows.

It was 1940 and the man making the request had arrived unexpectedly at the factory of the British printer and game publisher Waddingtons on 40 Wakefield Road, Leeds.

The company's workers thought it was an unusual request. The man said he was a businessman but he didn't want business cards. Instead he wanted visiting cards, the calling cards of high-society types who regarded the presence of contact details on their cards as common and rude. A visiting card would state a person's name and, at the very most, name the gentlemen's club they belonged to.

But business was business and it wasn't Waddingtons' place to second-guess a customer's needs, so they took the job. What name shall we put on the card? the employee taking the order asked. "E. D. Alston," the man replied.

When the visiting cards were ready, Alston returned, paid what he owed, and left. It seemed like just another printing job.

A few days later Alston returned and, with an air of authority, asked to meet Norman Watson, the head of the company, in private. Once alone, Alston told Watson that he was about to discuss a matter of national security and so the Official Secrets Act applied. Discussing what I am about to tell you with others can result in fines and imprisonment, Alston informed the shocked company chairman.

Alston was no businessman. Officially he was a civil servant handling the procurement of textiles at the Ministry of Supply offices in central Leeds, but that was just a cover. In reality he was a British intelligence agent working on behalf of MI9, a new branch of the British secret service founded on December 23, 1939, as the nation prepared for the imminent Nazi invasion of Western Europe. Its mission was to help combat personnel evade capture and help those taken prisoner escape.

MI9 was not the typical gathering of spooks. It was more of an ideas laboratory that invented new gizmos and schemes to get the troops out of enemy clutches. At the center of it all was technical director Christopher Clayton Hutton, an eccentric former movie publicist who was MI9's answer to James Bond's gadget guru Q.

He and his team designed pocket radios, made bars of soap with tiny compasses hidden inside, and constructed bicycle pumps that doubled as torches. They stitched together air force uniforms that could be turned into civilian clothing with ease and created waterproof provision tins filled with boiled sweets, water purification tablets, and slabs of brown, sticky "liver toffee," a combination of malt and cod liver oil that apparently tasted better than it sounds, which wouldn't be difficult.

Another idea to emerge from the secret think-tank in Room 242 of London's Metropole Hotel, which the British government had turned into offices in the run up to the war, was to smuggle silk escape maps into prisoner of war camps.

And that's where Waddingtons came in. For as well as being the publisher of Monopoly in the UK, Waddingtons was pretty much the only printer in the country who knew how to print on silk. Before the war, Waddingtons would print silk programs for the royals who attended the annual Royal Variety Command Performance, an evening of comedy, music, magic, and theater held in the name of charity.

Yet even for Waddingtons, printing on silk was a challenge. First the rayon, the artificial silk the company used, had to be treated with barium tungstate in order to make it opaque and enable double-sided printing. The process had to be done just right. Use too little barium tungstate and the silk wouldn't absorb the ink but use too much and the material would stick to the parts in the printing press.

After treatment the silk had to be mounted on the press. Again it had to be done carefully because even the slightest slackness in the material would cause the print to misalign. The ink also had to be specially mixed and, finally, the printed silk had to be oiled using special equipment to make it watertight.

Tricky as silk was to use, MI9 knew it was the perfect material for escape maps. Unlike paper, silk can be scrunched up and hidden in tiny spaces without incurring damage, making it easier to smuggle into prison camps and easier for servicemen to conceal. Silk doesn't rustle either so escapees could use the maps without the risk of giving away their position. On top of that silk maps are hard to tear, do not disintegrate in water, and, thanks to the oiling, the ink does not run or smudge in the rain.

MI9, Alston explained to Watson, wanted Waddingtons to print the hundreds of thousands of silk maps of Europe and Asia it planned to issue to British Commonwealth troops during the war. They also wanted the company to make Monopoly sets that doubled as escape kits that they could then send to British servicemen in the prisoner of war camps.

The Monopoly plan was typical of Clayton Hutton's maverick imagination. The cardboard base of the game board would have thin compartments cut into them so that a small compass, two files, and a silk map could be deposited inside. Then the paper with the playing area would be glued on top to conceal the equipment hidden in the board.

Specific property names on the board would be marked with a period so that officers inside the camps would know it contained escape equipment and MI9 would know which map it contained. A period on Mayfair, Monopoly's UK equivalent of Boardwalk, for example, signaled the inclusion of a map of Norway, Sweden, and northern Germany. One on Free Parking meant the game concealed a map of northern France and its borders with Germany.

Monopoly's bounty of paper money, meanwhile, would become a hiding place for Reichsmark bills and other Axis power currencies for prisoners to use as bribes or to aid their escape attempts.

Keen to support the war effort, Watson readily agreed to help and assigned three of the company's most trusted employees to the task. MI9 offered to provide armed guards to keep prying eyes away but Watson declined. It would only raise suspicion and frighten employees, he explained.

To maintain secrecy, Waddingtons put nothing down in writing about its MI9 work and began referring to Alston as Mr. A in case anyone not in the know overheard anything.

In the years that followed, Waddingtons produced hundreds of doctored Monopoly sets. In each one they secreted metal instruments supplied by MI9 and their own silk reproductions of the maps provided by the intelligence agency. Every so often, Watson would travel by train from Leeds to London and deliver the games to the lost luggage office in King's Cross railroad station. Later an MI9 agent would arrive at the office to reclaim the package.

Waddingtons was not the only game maker MI9 enlisted. Another recruit was John Jaques & Sons, the London company behind the Staunton chess set, which was tasked with producing chess sets and Snakes & Ladders games with hidden compartments for escape equipment.

With its supply of board games in place, MI9 turned its attention to getting them into the prisoner of war camps undetected.

Sending the games within British Red Cross aid parcels was out of the question. British prisoners relied on the food the Red Cross provided, and if a game containing escape equipment was discovered in one of the charity's parcels it would give the Axis powers grounds to stop accepting those life-saving supplies.

MI9's solution was to fabricate a national network of organizations with names like the Prisoners' Leisure Hours Fund, the British Local Ladies' Comforts Society, the Crown & Anchor Mission, and the Liverpool Service Men's Club. To give these nonexistent organizations legitimacy, MI9 co-opted the addresses of buildings that had been destroyed in Luftwaffe air raids and padded out its parcels with pages torn from the appropriate local newspapers.

For good measure MI9 sent letters from the dummy organizations to prison camp commanders to tell them they had raised money to send a parcel of games and other entertainments to the British inmates that would soon be on its way. Finally, MI9 sent coded letters from fictitious relatives to the British officers in charge of escape efforts in their prison camp so that they knew to look out for the parcel.

Once the parcels arrived, the escape officers would recover the hidden equipment and then burn what remained of the games in barrack stoves to hide the evidence. As such it's impossible to know exactly how many of MI9's games made it into the prisoner camps undetected but plenty did.

When the story of Monopoly's wartime activities went public in 1985, several former prisoners of war wrote to Waddingtons to confirm that they had seen the boards.

One of the veterans was J. T. Robson, a private captured in Greece in 1941. He spotted a silk map inside a damaged Monopoly board one evening while imprisoned at the Stalag XVIII-A camp near the Austrian town of Wolfsberg. He took his discovery to a superior officer who told him not to tell anyone else about it. Robson never found out what happened next, as shortly after the Nazis transferred him to another prison in the north of Austria.

The Monopoly games even made it into Colditz Castle, the infamous high-security prisoner of war camp where the Nazis sent troublesome Allied officers who had repeatedly tried escaping from other facilities. In a 1985 letter to Waddingtons, Bill Lawton told how he encountered several of MI9's Monopoly sets during his time as the

British assistant escape officer at Colditz. "Many of the items contained therein were invaluable to escapees from the castle, both successful and abortive," he wrote.

Not every parcel got through of course. MI9 reported that few of the parcels it sent to Italy got through, a failure it blamed on the "inefficiency of the Italian administration," which caused many of the parcels to be lost or delayed for months. The Germans, meanwhile, discovered how the British were using chess sets to smuggle items into its prisoner of war camps. The Nazis accused the British of breaching the Hague Conventions on the conduct of war and engaging in "gangster war." In retaliation they threatened to establish "death zones" around prisoner of war camps within which escapees would be shot dead on sight.

But most parcels got through. MI9 estimated that 90 percent of the twelve hundred parcels it sent out in 1941 arrived safely. Despite uncovering the escape tools hidden in chess sets, the Axis powers seemingly never suspected Monopoly's role in smuggling escape equipment in, and the Germans appear to have had little idea of the scale of the British smuggling operation.

By the end of the war just over 21,000 British Commonwealth personnel had escaped from prisoner-of-war camps and the UK government estimated that a third of those escapes were made possible by the activities of MI9 and the silk maps produced by Waddingtons.

The United States also used Monopoly to slip escape kits into the prison camps. Inspired by the activities of MI9, the Americans created an equivalent agency called MIS-X in October 1942.

Rather than working directly with Monopoly's U.S. owners Parker Brothers, MIS-X bought games from stores and brought them back to its base in Fort Hunt, Virginia. Its agents then carefully steamed off the paper from the board, chiseled thin compartments into the cardboard and filled them with silk maps, button-size compasses, files, and Gigli saws, the wire saws used by surgeons to cut through bone. The paper would then be reattached to the cardboard using the exact same glue as used in Parker Brothers' factory. Reichsmark bills were then added to the game's stock of paper money.

Once ready MIS-X, just like MI9, smuggled the games into prison camps under the banner of fake organizations and forewarned U.S. escape officers of their arrival with coded letters from pretend relatives.

While the British and Americans turned board games into escape kits, the Nazis used them for propaganda. After conquering France in 1940 the Nazis discovered that the world chess champion Alexander Alekhine was now a citizen of the Third Reich. When war broke out, the Russian-born grandmaster—who moved to Paris in 1920 and became a French citizen in 1927—had joined the French army as an interpreter, but when France fell he switched sides.

Encouraged, or possibly forced, by the Nazis, Alekhine began writing essays about chess that supported the Nazis' anti-Semitic ideology. In a series of six essays Alekhine decried the influence of Jewish players on the game and claimed that they played in a cowardly manner, preferring to seek a draw rather than a victory, unlike Aryan players who always strived to win.

After the war Alekhine tried to salvage his reputation by disowning his pro-Nazi writings and even claimed that someone else had written them. It didn't work and he was shut out of major chess tournaments for his Nazi connections. When he died in 1946, the original handwritten manuscripts of his pro-Nazi essays were found at his home.

But the Nazis' attempt to align chess with their abhorrent beliefs was amateurish compared to what followed, for the Soviet Union was about to take the idea of using board games as tools of propaganda and espionage to another level.

Nikolai Krylenko was the man responsible for cementing chess within Soviet life. After briefly heading the Red Army, he became the Soviet Union's chief prosecutor and used his position to promote the concept of "socialist legality," where political considerations trumped evidence, decided guilt, and dictated the punishment.

When not occupied with show trials, Krylenko banged the drum for chess, which he regarded "a scientific weapon in the battle on the cultural front." For the communists, chess embodied the logic and rational ideals they wanted to instill in the masses to inoculate them against religious belief. The Bolsheviks also thought international chess tournaments would provide a vivid way to demonstrate to the entire world that the Soviet system was intellectually superior to capitalism.

Encouraged by Krylenko, the communists promoted chess as an integral part of Soviet culture. Chess clubs opened across the USSR, springing up in factories, schools, military barracks, and collective farms. The state also sponsored the careers of promising players. Sales of chess magazines boomed and by 1929 around 150,000 players were enrolled in the Soviet chess program.

Krylenko wanted more though. In 1932 he told a gathering of chess players that even more must be done to make chess a tool of communism. "We must organize shock-brigades of chess-players, and begin the immediate realization of a five-year plan for chess," he declared.

His call seemed like yet another manifestation of the Soviets' desire to hammer away at reality until it matched their ideology, up there with demands for Marxist mathematics and Leninist physics.

But even as he called for a five-year plan for chess, Krylenko's mission to turn the game into a cultural weapon was gathering pace. By 1934 a half million people had joined the state chess program and estimates suggested that half of the world's chess players lived in the USSR. What the Soviets needed next was a champion. A communist grandmaster who could go forth and deliver a victory over the capitalists on the cultural battlefield of the chessboard. And just as in the warped world of socialist legality, the end would justify the means.

The first test of the Soviet chess machine came at a USA versus USSR tournament held on September 1, 1945. The Second World War had ended just days before and this contest was the first international team sport event the Soviets had agreed to compete in. It was also to be played remotely with each move sent between New York and Moscow via radio telegraphy.

The opening game saw U.S. champion Arnold Denker facing Mikhail Botvinnik, the Soviets' foremost player. It was a fiery game that had the hundreds of spectators who crammed into the ballroom of the Henry Hudson Hotel in Manhattan on the edge of their seats.

Botvinnik's black pawns charged out daringly from the center in the opening stages only for the momentum to then swing in Denker's favor. Soon it looked as if the American had Botvinnik on the ropes with the Soviet contender's pieces hemmed into his queen side of the board. Then, to the shock of the audience, Botvinnik flipped the game on its head. One moment Botvinnik was on the back foot and then in the space of just five moves he ripped apart his opponent's army, capturing Denker's queen, cornering his king and forcing the American champion to concede defeat.

More Soviet victories followed. When the tournament ended on September 4, the USSR had inflicted defeat after defeat on the U.S. team, which managed just two wins in the twenty games played. This wasn't a contest, it was a massacre.

The Soviet chess machine was just getting started. Three years later Botvinnik headed to The Hague for the first stages of the World Chess Championship. The USSR's odds were good; three of the five grandmasters in the contest were Soviets.

Despite this the Soviet chess officials worried that the United States' Samuel Reshevsky would win so they tried to rig the tournament. The officials began leaning on Soviet player Paul Keres, dropping hints about how he should lose his games against Botvinnik for the good of the nation. By throwing the games, Keres would hand Botvinnik enough victories to ensure that Reshevsky couldn't win the tournament.

Keres was certainly vulnerable to the pressure. He was an Estonian and the USSR was in the midst of a military campaign against the Estonian resistance. The authorities also suspected Keres of collaborating with the Nazis when they occupied Estonia during the war. It wouldn't take much for the Soviets to find a premise to freeze him out of international tournaments, or worse, if he didn't cooperate.

Whether Keres threw the game is unknown. But even if Keres did not comply, the Soviet authorities got what they wanted. Botvinnik's victories over Keres gave him the very points that he needed to outscore Reshevsky and be crowned world chess champion.

Botvinnik's victory in 1948 ushered in an era of Soviet dominance in chess. For the next twenty-four years the USSR held the world title. The Soviet chess program, Botvinnik declared in 1949, had turned the nation's players "from 'pupils' into teachers." As the USSR cemented its status as the motherland of chess, it began promoting the idea of the Soviet School of Chess, a "scientific" approach to the game characterized by careful pre-match preparation and the development of new, bold methods of play.

The victories delivered by the Soviet grandmasters became a source of national pride within the USSR and useful fuel for the Cold War propaganda it exported. Chess even became a handy way for Soviet spies to exchange messages. The KGB developed a code that allowed moves in what appeared to be unremarkable play-by-mail games of chess to convey instructions to agents in the West and for spies to send updates back to Moscow.

The Soviet chess machine seemed unstoppable, but then rumors began circulating about a young chess wizard from Brooklyn: Bobby Fischer. After joining a chess club at the age of six, he'd become obsessed with the game and constantly pestered his mother to take him to Washington Square Park to play. By the middle of the 1950s Fischer was playing against masters, and when he became the U.S. chess champion in 1958 he was just fourteen years old.

It seemed inevitable that this nail-biting kid would one day play for the world title. Provided he didn't self-destruct first, that is. For though Fischer was an exceptional chess player, he was also an arrogant prima donna.

His prickly personality was a constant source of trouble. In 1958 the Soviets invited him on an all-expenses-paid trip to Moscow, the chess capital of the world. Once there Fischer bewildered his hosts by refusing offers to see sights like the Kremlin because he only wanted to play chess. He followed that up by demanding to know how much the Soviets would be paying him to play against their grandmasters—nothing was the answer—and calling the Russians "a bunch of pigs" while at the Moscow Chess Club.

Back in the United States he poured scorn on everyone and everything, including well-wishers. When the U.S. women's chess champion Lisa Lane called him the greatest player who ever lived in 1962, Fischer told *Harper's* magazine: "The statement is accurate, but Lisa Lane really wouldn't be in a position to know. They're all weak, all women. They're stupid compared to men."

At other times he raged against Jews, the United States, his mother, homosexuals, the rich, the poor, teachers, the Kennedys, and, well, just about everything else. The only things he seemed to approve of were his own tailor-made suits and chess, the game that obsessed him so much that one of his ambitions was to live in a house built in the shape of a rook.

But for all his personality flaws, Fischer remained the only man who seemed capable of breaking the communist stranglehold on chess. Not that the USSR intended to make it easy for him. Soviet players were now backed by large teams who would provide advice on how to secure victory during breaks in international games. The idea of rigging international matches had also taken root with Soviet players, as they planned out their matches in advance to improve the chances of victory for the state's preferred grandmasters.

By the end of the 1960s the twenty-six-year-old Fischer finally seemed ready to try for the world title. The only trouble was making sure the hotheaded grandmaster would participate, as he had now taken to walking out of tournaments or refusing to play if conditions were not to his liking.

None of which helped those trying to organize the 1972 showdown between him and then world champion Boris Spassky in the Icelandic capital Reykjavík.

By the time Fischer was finally persuaded to play, the looming showdown between the superpowers' chess champions had captured the world's imagination. The clash was seen as a microcosm of the Cold War; the global struggle between East and West reduced down to two men battling on a chessboard.

Not that either player embodied their homeland's ideals. The vain and vulgar Fischer hardly personified how the United States saw itself; Spassky was an unashamed dissident who had turned against communism after witnessing the Soviet crackdown on Czechoslovakia in 1968 and was only tolerated because of his mastery of chess.

Fischer started badly. He lost the first game and then lost another without making a single move because of an argument about resetting the chess clocks. But then Fischer pulled himself together and began scoring victory after victory. By the tenth game he had a sizable lead and the Soviets were getting twitchy.

The Soviets publicly accused the United States of using chemicals, computers, and other methods to undermine their man. In response the match was put on hold so the Soviets could search the hall with X-ray machines, metal detectors, and Geiger counters. They found nothing but remained paranoid. Maybe the CIA had tampered with Spassky's food or were using a parapsychologist to interfere with his thoughts. Maybe a spy had infiltrated their chess team and passed details of Spassky's pre-match preparations to the Americans.

Fischer was no less paranoid. He convinced himself that the KGB might assassinate him or put something in his food and drink or hypnotize him. To try and protect himself from these threats he demanded that the orange juice he drank be made fresh from oranges squeezed in front of him and asked for round-the-clock protection from the Marines, a request that was refused.

After the Soviet sweep of the tournament venue, the match resumed, and when Fischer defeated Spassky in the twenty-first game, he became the first non-Soviet world chess champion in twenty-four years. American newspapers and magazines hailed Fischer as a hero and interest in the game revived in the West, with a rush of new books and soaring sales of chess sets. Fischer celebrated by boasting how he had single-handedly defeated the Soviet chess machine.

But Fischer was a blip. After his victory in Reykjavík, he got into a spat with the global chess organization FIDE over the terms of the next world championship contest and refused to defend his title unless he got his way. FIDE eventually lost patience and in 1975 decided that by refusing to play, Fischer had forfeited his title to the Soviet grandmaster Anatoly Karpov, the man the American would have played if he had agreed to play.

Fischer never fought for the title again and became a recluse who obsessed about the KGB coming after him. He would keep a fruit juice squeezer in a locked suitcase to prevent communist agents tampering with his food and drink, and even got his dental fillings removed because he thought they could be used to influence his thoughts. By the time he died in 2008, Fischer had fled to Iceland to avoid arrest in the United States for tax evasion and become a conspiracy theorist and Holocaust denier who declared that America "got what it deserved" on September 11, 2001.

While Fischer exiled himself from chess, the Soviets soon faced a new challenge to their grip on chess. Only this time the challenger was one of their own.

Viktor Korchnoi had been one of the USSR's most promising chess players, a Leningrad-born grandmaster who had won the USSR chess championship four times. But the authorities felt he was politically suspect and focused their efforts on helping Karpov instead.

In 1976, fed up with the suspicion and restrictions placed on him, Korchnoi claimed asylum while in Amsterdam and eventually settled in Switzerland. The Soviets were livid and launched a smear campaign against him. At home the Soviet newspapers branded him a traitor and renegade.

But when it became clear that Karpov would be defending his title against the defector at the 1978 world chess championship in the Philippines, the KGB decided it had to do more than hurl insults. To put pressure on Korchnoi the Soviets ordered his son Igor, who was still in the USSR, to be drafted into the army, a move that would bar him from leaving the country for five years. Igor went into hiding to avoid the draft but was eventually caught and imprisoned in June 1978. The KGB also sent a squad of agents to the championship ranging from bodyguards to people who would scan Karpov's lodgings for bugs and monitor his health. There was even a parapsychologist tasked with trying to undermine Korchnoi during the game using telepathy.

Korchnoi's challenge failed and Karpov remained the champion, but in 1981 the two went head-to-head once again at the subsequent world championship in Merano, Italy.

Worried that Karpov's lodgings would be bugged, the Soviets brought with them a huge shielded tent that prevented radio waves from entering or leaving so that the grandmaster could prepare for games without the risk of people listening in. For good measure Karpov was put under constant guard and all outsiders were banned from visiting him.

In the book *The KGB Plays Chess,* former KGB operative Vladimir Popov claimed that if Karpov looked likely to lose, Soviet agents planned to assassinate Korchnoi by administering a toxin that would cause immediate heart failure. If the poison existed it wasn't needed, for Korchnoi lost his second and last attempt to take the world title from Karpov. The Soviet chess machine had triumphed once more.

After that Karpov and later Garry Kasparov ensured that the USSR would maintain the world title right up to the moment that the communist superpower was finally dissolved in 1991 following the end of the Cold War.

With the exception of Fischer's brief reign as world champion, the

USSR had dominated the chess world for forty-three years. Krylenko, who became a victim of his own socialist legality in 1938, got the "shock-brigades" of Soviet grandmasters he wanted, but the communist regime he imagined them protecting fell apart all the same.

Chess, it turned out, was just a game after all.

8

# CLUE'S BILLION-DOLLAR CRIME SPREE

## *How Clue's very British murders created a world of armchair sleuths*

Birmingham is burning. It is early 1943 and the powerhouse of British industry is under attack from the German Luftwaffe.

At night air raid sirens scream and the citizens race to the safety of shelters as bombers drone overhead and antiaircraft guns thunder. In the morning there is smoke and the rubble of another night of war to clear before darkness falls and the cycle starts over.

As the city licks its latest wounds, inside the factories workers busy themselves with the task of making bullets, bombs, guns, and the other military necessities of war. In one of these factories is Anthony Pratt. Deemed unfit for military service because of his poor eyesight, he now spends most of his waking hours at the C.O. Ericsson Engineering Works in Kings Norton. Before the war the factory made kitchen appliances. Now it builds parts for tanks.

Pratt is doing his part for the war effort, including shifts as a fire warden, but he's bored. Life in the factory is tedious and the outside world not much better. On the dinner table each evening, the dull, depressing food ration. At night nothing to do but stay indoors listening to the constant stream of wartime propaganda on the wireless.

Outside his car lies idle, out of action because gas is reserved for essential travel. Not that there's anywhere to go now. In the evenings the neighborhood is empty and black, the street lamps switched off to save power and hide the city from the danger in the skies.

It's a life of dreariness, where the only breaks are the bursts of terror that come when the bombs rain down. And then there's the increasing rudeness of people and the sloppy service in the stores, all excused with the irritating mantra: "Don't you know there's a war on?"

It was all so different from his life before the war. That was the life. A life of fun and adventure.

Before the war Pratt was a pianist and composer. He performed in lavish country hotels, tinkled the ivories on luxury cruises to New York, and even played alongside the famous Norwegian opera singer Kirsten Flagstad. And even when his musical career stalled and he became a solicitor's clerk, he still got to enjoy himself at the parties his friends threw.

"We lived like lords," he recalled years later. "Then came the war and the blackout and it all went, 'Pouf!' Overnight all the fun ended. We were reduced to creeping off to the cinema between air raids to watch thrillers."

At least he still had his books. The ghost stories, the Charles Dickens novels, and, of course, the murder mysteries and accounts of real-life crime that he loved. Well, at least until the sirens whirred into life and the lights had to go out again.

The tedium of life during wartime did, however, leave Pratt with plenty of time to contemplate ways to brighten things up, and one day he decided to make a board game.

His friend Geoffrey Bull had made one. Buccaneer it was called: a tabletop adventure set in the Caribbean Sea where players raced to plunder bounty from a treasure-filled island and take their haul back to port before a rival player could raid their ship and steal the cargo. The game had even been published by Waddingtons, the Leeds company that had brought Monopoly to Britain.

But what kind of game to make? A game of murder, of course. A game that would bring his beloved murder mysteries to life. A game that could recapture the thrill of "murder," the parlor game played at all those prewar parties, where guests had to work out who among them was the murderer before he or she killed them with a tap on the shoulder or a wink.

With the theme in place, Pratt and his wife, Elva, immersed themselves in creating a game where players competed to be the sleuth who solved the murder. Pratt found inspiration on his bookshelf, filled as it was with the grand British murder mysteries of the so-called Golden Age of Detective Fiction that had taken the country and the world by storm in the years leading up to the Second World War.

The roots of the golden age go back to the early nineteenth century when the Victorians turned murder into a thriving entertainment industry. As Britain urbanized and state-funded schools improved literacy rates, the nation began devouring the cheap and sensationalist broadsides: single-sheet newspapers that offered sordid and detailed accounts of real-life murders. They sold millions and were often hawked to the crowds at public hangings so that, having watched the murderer meet his end, they could enjoy reading every detail of the crime at leisure.

The often-embellished tales of the broadsides were only the start. Murder scenes became destinations for curiosity seekers, and police had to search for evidence amid crowds of gawkers who hoped to see a fresh corpse or bloodied murder weapon. Others bought trinkets commemorating famous murders or watched puppet shows that reenacted real-life homicides.

By the 1860s the British public's tastes started to become more refined. Instead of stories about murderous lowlifes, they now wanted tales of murder and crime in the respectable circles of the upper and middle classes. Wilkie Collins was the king of these new "sensation novels" and his 1868 book *The Moonstone* set the blueprint for British murder mysteries, even though it concerned the theft of a valuable jewel rather than a homicide.

*The Moonstone* took place in a large country house where all the wealthy suspects, each with secrets and scandals to hide, had gathered for a party. A detective had arrived to investigate the crime and the story contained hidden clues to the identity of the criminal so that readers could try and solve the mystery before they reached the end where all would be revealed.

From there the path to the Golden Age of Detective Fiction was straightforward. As the years went by the investigator became more important than the crime, leading to detective heroes like Arthur Conan Doyle's Sherlock Holmes.

By the 1930s the transition was complete. The gruesome detail of the broadsides had been cast aside and the task of solving the crime had taken center stage. Where once the British lapped up stories detailing every slash and stab of the killer's knife, they now turned to novels where murder was tidy and sanitized. Victims died peacefully rather than painfully from poisons. Gunshot wounds never made a mess. Sometimes the body had even been removed from the scene of the crime before the story began.

The characters in these books often seemed just as removed from the grim reality of murder as the reader, acting as if the death was just an unfortunate event on par with being caught in an unexpected thunderstorm without an umbrella.

Even the detectives had been tamed. Instead of the unpredictable, opium-addled Holmes, the golden age offered the impeccably dressed and punctual Hercule Poirot and the always-polite Miss Marple. Now the mystery was everything and readers loved trying to solve the crime as they read, before the culprit, after the inevitable twists and red herrings, was unmasked by the respectable detective.

Such was the appeal of these stories that in 1934 one in eight of all books published in Britain was a crime novel. The stories of Edgar Wallace, Dorothy Sayers, G. K. Chesterton, and—above all—Agatha Christie sold in enormous numbers not just in Britain but throughout the world.

The work of Christie and other British detective novelists was a world apart from the books being written on the other side of the Atlantic. In the United States authors like Raymond Chandler drew on the country's history of outlaws and gangsters to deliver urban tales of hard-boiled and hard-drinking detectives. Yet even in the United States, the starchy-yet-cozy British tales of gentile murders in rural country houses sold in huge numbers.

The country house where many, but far from all, golden-age detective novels took place played a big role in giving these books global appeal. In both the UK and beyond the British country house has become an icon. Even today, when most country houses in Britain are no more than tourist attractions or dilapidated wrecks housing dusty lords who can't afford to maintain them, the idea of the country house resonates across the world.

Just look at *Downton Abbey,* the British soap opera about life in an early twentieth-century country house that attracted an audience of around 120 million viewers across 200 countries.

The British country house is a product of politics. Unlike most of eighteenth-century Europe, Britain did not have an absolute monarch but one kept in line by the nation's Parliament. As a result the rich and powerful did not need to hang around the royal court; they could simply go to London when Parliament was in session and spend most of their time on their sprawling estates.

These estates soon became showcases for their owners' wealth and power. The houses were decorated with fine fabrics, marble floors, and ornate staircases, and manned by squads of servants. The land around them, meanwhile, was transformed into a picturesque landscape designed for residents and guests of the country house to gaze at through the enormous windows as if it were a painting.

While the country houses were about grand displays of wealth and political influence, they seeped into the very fabric of what Britain thought of itself and the world thought of Britain. The country house came to represent something benign and comforting, an escape from the realities of an uncertain world. A place of continuity and safety that would endure in the imagination long after the sun set on the British Empire.

And it was this image that the British detective novelists toyed with. After all where better to stage a shocking murder than within the splendid isolation of these tranquil country houses?

So the Pratts' decision to set their murder mystery game within a plush country house was a no-brainer. They named it Tudor Close, and Elva created a board that mapped out the ten rooms on its ground floor: the lounge, the dining room, the kitchen, the ballroom, the study, the library, the billiards room, the conservatory, the gun room, and, in the middle of them, the cellar.

Next came the guests; a gathering of detective fiction archetypes named after the color of their playing piece: Dr. Black, Mr. Brown, Mr. Gold, the Reverend Green, Miss Grey, Professor Plum, Miss Scarlet, Mrs. Silver, Nurse White, and Colonel Yellow. At the start of every game one of these characters would be randomly selected to become the murder victim. The surviving guests became both the investigators and the suspects.

Finally there were the murder weapons: an axe, a cudgel, a fireplace poker, a pistol, a dagger, a length of rope, a bomb, a hypodermic syringe, and a bottle of deadly poison.

After selecting the murder victim, cards representing the murderer, the weapon, and the room that was the scene of the crime would be drawn from the packs unseen and placed in an envelope. The players then had to deduce what was inside the envelope by collecting the remaining cards during visits to the various rooms of Tudor Close and testing their theories to see if other players could disprove their suspicions with the cards they had. Once a player believed they had solved the mystery they would make a final accusation and check the envelope to see if they were right. If they got it right they won; if not they were eliminated from the game.

The Pratts spent months refining the game and testing it on their friends, who encouraged them to keep working on it whenever they began to lose interest. By December 1944 the game, which they named Murder, was finished and Pratt applied for a patent. Then his game-making friend Geoffrey Bull arranged a February 1945 meeting with

Waddingtons so the Pratts could show their game to Norman Watson, the company's managing director.

Watson was impressed and agreed to publish it, but insisted there had to be changes.

First the name. Calling the game Murder, said Watson, was "most unsuitable." So Waddingtons came up with a new title: Cluedo, a fusion of the words "clue" and "*ludo,*" the Latin for "I play." The gun room. It also had to go and was replaced by an extension of the dining room. Secret passageways connecting different rooms were added to make it easier for players to travel around the board.

Waddingtons also felt there were too many characters and so it slimmed down the guest list. Mr. Brown, Mr. Gold, Miss Grey, and Mrs. Silver had their invitations to the party at Tudor Close withdrawn. Nurse White was demoted to a cook and renamed Mrs. White. A new character, the elderly grande dame Mrs. Peacock, joined the line up, and Colonel Yellow became Colonel Mustard so as not to associate a military man with cowardice. Dr. Black became the owner of Tudor Close and the game's permanent murder victim. Only the Reverend Green, the femme fatal Miss Scarlet, and the scatterbrained Professor Plum survived unchanged.

Finally Waddingtons revised the selection of murder weapons. The pistol got upgraded to a revolver while the axe, the cudgel, the bomb, the poison bottle, and the syringe were discarded. In their place were a spanner, a candlestick, and a section of lead piping.

Revisions complete, Waddingtons was ready to publish, but even once the Second World War ended material shortages continued, forcing Cluedo's release to be delayed by years. When Cluedo finally went on sale in the UK in late 1949, it was already out in the United States thanks to Parker Brothers, the game publisher that had given Waddingtons the license to produce Monopoly in Britain and beyond.

When Waddingtons first suggested that Parker Brothers might want to release its murder mystery game in North America, the company hadn't been interested. "It so happens that our founder,

Mr. George S. Parker, has laid down as a policy of the company an unbroken rule that we will not publish any games dealing with murder," Parker Brothers' president Robert Barton informed Watson by letter in May 1948. "'Cluedo,' from what little I have played it, seems to be a most excellent game, but unfortunately we cannot publish it because of this rule."

But the more Barton played Cluedo, the more convinced he became that he should push the company's founder to let him break the rule. In August 1948, Barton told Watson: "After all, London and Scotland Yard have always been considered in the English-speaking world as the home of the great mystery and detective stories. An English game on such a subject should be a natural success over here."

Eventually, Barton persuaded Parker to let him break the policy but the company felt the game needed further changes before being released in the United States. Parker Brothers renamed the game Clue and gave it the subtitle: "The Great New Detective Game." The dagger became a knife and the spanner a wrench. Tudor Close was renamed Tudor Mansion and Dr. Black became Mr. Boddy, a joke name inspired by his status as the game's perpetual murder victim.

The rulebook, meanwhile, was tweaked to reflect the company's squeamishness about the theme with mentions of "the murder" changed to "the act." Finally Reverend Green was defrocked to become Mr. Green since Parker Brothers worried that having a member of the clergy cast as a potential murderer would not go down well with the American public.

Even after all this, Parker Brothers remained uneasy about publishing Clue. As its release approached, Barton informed Waddingtons that the company would not be advertising the game in magazines to coincide with its launch because they were still concerned about being associated with a game about murder. The company, he added, expected that another game it had bought from Waddingtons—a Parcheesi clone called Skudo—was more likely to be a success because it had a lower retail price and fitted the mold of games already popular in the US. At most Parker Brothers hoped Clue would break even, regarding its publication more as a favor to Waddingtons than a wise business move.

At first Parker Brothers' doubts seemed correct. Its salesmen reported that retailers were opting for Skudo over Clue, but the tables soon turned. Clue's sales quickly outstripped those of Skudo and it became Parker Brothers' top new game of 1949 with thirteen thousand copies sold. By early March 1950 another ten thousand had been sold and Parker Brothers decided to pay Arthur Conan Doyle's estate a large sum of money so it could promote the game as "The Great New Sherlock Holmes Game."

The story was much the same in Britain. Initially the game sold below Waddingtons' expectations but then sales began growing steadily as excitement about it spread through word of mouth.

As the 1950s wore on it became clear to Waddingtons and Parker Brothers that, despite the initial disappointment, the Pratts' game was still gathering steam, with sales rising year on year and showing no sign of stopping. Even by the end of the decade its annual sales were still rising.

The Pratts made relatively little money from its success. In May 1953 Waddingtons persuaded them to accept a one-off payment of five thousand British pounds for the rights outside the UK. It was, for the time, a huge sum, almost equivalent to 190,000 dollars in today's money. Plus, the Pratts would still continue earning royalties from sales of the game in Britain until the patent expired in 1967.

By the time the patent ran out the game had become one of the world's bestselling board games, but Waddingtons did not even bother to write to Anthony Pratt to acknowledge that their relationship was over. Eventually Pratt had to write to Watson to say that he assumed the deal was over and to thank Waddingtons for believing in his game in the first place: "I have a certain wistfulness at the prospect of relinquishing such a handsome income, more especially as it was still ascending at the moment of cessation, but I have much to be thankful for."

By 1975 the game's success had grown even bigger with around two million copies being sold worldwide every year. While Skudo was quickly forgotten, Clue joined the ranks of the world's favorite board games, alongside the likes of Monopoly, the Game of Life, and Scrabble. Only, unlike the authors of those games, Pratt missed out on the fortune he could have made.

Pratt never had another game published. In public he shrugged off the fortune he missed out on, but according to his daughter Marcia, behind closed doors they—and Elva especially—felt they had been swindled. Much of the money they made from the game was eaten away by rampant inflation and when it ran out Pratt returned to a life of office work and faded into obscurity. The world knew his game but almost no one knew his name.

In 1996 as total sales of the game passed the 150 million mark, Waddingtons launched a phone hotline to try to find Pratt as a publicity stunt. They soon discovered that Pratt had died two years earlier in a care home after developing Alzheimer's. His wife, Elva, had passed away in 1990.

For Waddingtons, the Pratts' game was more than a lucrative hit. It also prevented Parker Brothers from revoking its license to produce Monopoly in the UK. The unspoken threat was that if Parker Brothers canceled the Monopoly license, then Waddingtons would retaliate by removing Parker Brothers' right to make Clue in the United States. Hasbro, the company that bought Parker Brothers in 1991, eventually solved the problem by spending fifty million pounds—around seventy-five million U.S. dollars—to take over Waddingtons in 1994.

Clue's appeal also went beyond the confines of the tabletop. Unlike almost every other board game, Clue's famous characters and murder-mystery theme made it perfect for spin-offs. The game spawned novels, puzzle books, and even KFC Kids Laptop Meals that gave children Clue-themed brainteasers to solve alongside their fried chicken strips and mashed potato.

In 1985 the game became a comedy film that offered cinema-goers three versions, each with different endings. That same year the *Clue*

*VCR* game, which sought to bridge the gap between watching videos and playing the game, became one of the bestselling videocassettes in the United States alongside *Ghostbusters* and *Beverly Hills Cop*. Television series followed, including a five-part 2011 miniseries for TV that recast the familiar characters as a group of teenage investigators. Clue fans could even watch a stage musical based on the game in the mid-1990s and buy a miniature dollhouse based on Tudor Mansion.

Like the country houses that helped inspire it, Clue had become an enduring icon. The image of monocled Colonel Mustard and Professor Plum with his "Doc" Brown from *Back to the Future* hair seared themselves on the world's consciousness just like Holmes, Poirot, and Miss Marple before them. In many ways Clue, despite not being a novel, came to be the ultimate embodiment of the Golden Age of Detective Fiction and its archetypes.

But with the 1920s setting so integral to its identity, keeping Clue in tune with the times has proven difficult.

Waddingtons tried first. In 1986 the company released a more complex spin-off called Super Cluedo Challenge. Created at Waddingtons' request by freelance game designers Malcolm Goldsmith and Michael Kindred, Super Cluedo Challenge surprised the company. Waddingtons had expected the pair would merely add some more characters, rooms, and weapons. Instead they got that plus a game that tried to add a new dimension to the detective action.

In the original game players solved the mystery through a process of elimination, ruling out rooms, weapons, and suspects until the answer became clear. Super Cluedo Challenge added the chance to solve the mystery by gathering evidence too. Rather than being placed inside a single envelope, the murderer, weapon, and location cards now went into separate plastic holders with flaps in each corner. When a player landed on a clue counter they could get the chance to look under a flap to see part of a code that could be used to identify which card was inside the holder.

Even so Waddingtons felt the new game wasn't as good as the original. In a company memo, its chairman Victor Watson expressed concern that the new version detracted from the player interaction that was central to the original game.

It would be another twenty-two years before the next major revision of the game was attempted with 2008's Clue: Discover the Secrets. While Super Cluedo Challenge fiddled with how the game played, it had left the country house setting in place. But with Discover the Secrets, Hasbro not only changed the game but transported it into the present day.

Mr. Boddy's old-fashioned mansion was bulldozed and rebuilt as a flashy Hollywood Hills pad with a spa, swimming pool, movie theater, and observatory rather than anything as antiquated as a library or billiards room. Instead of a party for fusty society types, the mansion was now the scene of a paparazzi-attracting gathering of celebrities and tycoons.

Among the star-studded guests were American football star Jack Mustard, the movie star Kasandra Scarlet, Hollywood's slickest agent Jacob Green, and the video game billionaire Victor Plum. And when these guests needed a murder weapon they turned to baseball bats, awards trophies, and dumbbells rather than a wrench or lead pipe.

The game changed too, most notably with the addition of the "Clock Cards" that lay in wait within the deck of cards. Any player that drew one of these cards would become the latest murder victim and be eliminated from the game. These celebrities were no longer up against some gentile British murderer but a serial killer, and if the players failed to solve the mystery before the last clock card was pulled from the deck, the psychopath would have slaughtered all of them.

Hasbro called Discover the Secrets a "much-needed facelift" for the game, but Clue traditionalists hated it for throwing out the familiar trappings of the Golden Age of Detective Fiction. How could they do that to Miss Scarlet, let alone poor old Professor Plum?

In the end the traditionalists won out. Discover the Secrets is no longer in production.

After being thwarted in its bid to bring Clue into the present day, Hasbro changed tack and in 2016 injected modernity into the game's

1920s setting by killing off Mrs. White. To replace the downtrodden servant they added the much up-to-date Dr. Orchid, the adopted daughter of Dr. Black and an expert in plant toxicology who appeared to be of Asian descent.

And yet even as Mrs. White became the latest victim of Tudor Mansion, the period setting of the game remained unchanged. Just like the British country house, Clue, it seems, has implanted itself in our imaginations as an enduring and timeless escape from the modern world.

9

# SCRABBLE: WORDS WITHOUT MEANING

*Why words are meaningless to the best Scrabble players*

"Necessity is the mother of invention," and that was certainly true for Alfred Butts.

When the stock market went into free fall in 1929 he was a draftsman designing suburban homes for a New York architectural practice. Two years later he was laid off and had joined the growing ranks of the unemployed—yet another victim of the Great Depression.

Unable to find a new job, the mild-mannered and bespectacled Butts spent his days at his apartment in Jackson Heights, Queens, thinking of ways to pay the rent. He wrote magazine articles, penned stage plays, painted watercolors, and tried selling prints of famous New York scenes but he still found his pockets empty.

Another money-making idea he had was to create a game. Butts had always enjoyed chess; he had even been a member of the University of Pennsylvania's chess team while studying for his architecture degree. But what kind of game to make?

Butts was a meticulous man, someone who preferred to probe the evidence for answers rather than rely on the inventiveness of his own imagination. As such his instinct was to answer the question of what game to make by studying the games already on sale. Following his research he decided there were three types of game.

First there were "move games" that involved moving pieces around the board and counted chess and checkers among their number. Next, there were the "number games" like bingo and dice where the goal was to get the right figures. Finally there were "word games" like the newspaper anagram puzzles that he enjoyed solving.

He thought a game of skill such as chess would be too highbrow to sell to the mass market but since he never enjoyed the randomness of dice, he concluded that his game should be half luck, half skill.

Butts settled on an anagram game where players would pull letters at random and then use them to construct words. Almost immediately he ran into a problem: how could he minimize the chance of players ending up with a random assortment of consonants that prevented them from forming a word?

He found the answer to his conundrum in the pages of Edgar Allen Poe's 1843 short story *The Gold-Bug*. In the tale the hero uncovers a stash of buried treasure after decrypting a message written in a code based on how often each letter of the alphabet appears in the English language. Inspired by Poe's code, Butts decided to make the mix of letter tiles in his game correspond to how commonly those letters are used in English. To get the balance right, the ever-meticulous Butts began going through newspaper article after newspaper article, recording how often each letter appeared in words of differing length.

Butts was far from the first to turn to newspapers when trying to estimate letter frequencies in English. When U.S. inventor Alfred Vail began developing Morse Code he decided his alphabet of dots and dashes should be designed so that common letters took less time to tap out on telegraph machines. So Vail headed to his local newspaper office where he calculated the frequency of letters by counting the number of letters in the newspaper's type case.

Butts's more labor-intensive approach meant spending weeks pouring over the pages of the *New York Herald Tribune, The Saturday Evening Post,* and more. His hands smudged with ink, he sampled word after word from the articles, carefully counting how often each letter appeared and entering the results into an ever-growing pile of paper spreadsheets. His letter counts were nothing if not thorough. In one of his many trawls through the newsprint he counted 6,083 letters from *The New York Times* alone. Many would find this a tedious test of patience but Butts liked things organized and orderly. He had even diligently indexed his sizable postcard collection. After completing his surveys, Butts calculated the frequencies of each letter and let the results dictate how many of the hundred tiles in his game should represent each letter of the alphabet.

Butts called his word game Lexiko and the goal was to make nine- or ten-letter words using letter tiles picked at random. He pitched Lexiko to game publishers but none of them wanted it. Undeterred Butts opted to sell the game himself. He sat down at his kitchen table in late 1933 and began gluing letters onto plywood squares and fashioning racks that would make it easier for players to rearrange their letters into winning words. By August 1934 he had sold just eighty-four and the cost of making them had left him twenty dollars poorer than when he began. Butts concluded that Lexiko simply wasn't good enough and needed improvement.

The first addition to the game was a board and Butts found inspiration for how it should work in newspaper crossword puzzles.

Arthur Wynne, a Liverpool-born journalist who left Britain for the United States in 1891 at the age of nineteen, was the inventor of these brainteasers. In 1913 he was the editor of the *Fun* supplement that appeared in the *New York World* newspaper every Sunday and provided readers with a collection of weekly mind challenges from anagrams to rebus puzzles.

As the year drew to a close, Wynne's bosses began pressing him to come up with a new "mental exercise" to keep things fresh for readers. Racking his mind for ideas, Wynne thought about a game called "magic squares" that his grandfather taught him back in England.

Magic squares confronted players with an empty grid of squares and a set of words. The players then had to write letters into the squares on the grid in such a way as to ensure the list of words was spelled out correctly when you read down and across the grid.

Wynne took magic squares' words-on-a-grid format and replaced the word list with hints about what the words were. He dubbed his new puzzle Word-Cross and it made its debut in the December 21, 1913, edition of the *New York World*.

Readers who turned to the *Fun* section that day found themselves presented with a diamond-shaped grid with an empty middle and a simple instruction: "Fill in small squares with words which agree with the following definitions."

Some of the clues were simple but others were obscure or cryptic. The fiber of the gomuti plant, posed one clue. Doh, apparently. What we all should be, stated another, leaving readers to deduce that the answer was "moral." No wonder one wry clue asked what this puzzle is. The answer? "Hard."

Word-Cross didn't last long. A few weeks later one of the newspaper's typesetters got mixed up and changed its name to Cross-Word. It didn't matter because Cross-Word became an instant favorite among readers. Whenever the *Fun* section failed to include it, people would write letters of complaint.

Thinking he had hit on something special, Wynne urged his editors to copyright or patent his puzzle but they refused. Cross-Word was a pain to put together and the fickle readers would tire of it any day now, Wynne's bosses reasoned. Other newspaper editors agreed and when Wynne retired in 1918 the crossword puzzle was still something only found in the *New York World*.

But that would all change when Richard Simon went for dinner at his aunt's in 1924. Over dinner the aunt asked Simon, who was trying to get into book publishing, if there was a collection of Cross-Word puzzles she could buy for her daughter. No such book exists, Simon told her before thinking: "Why don't I publish one?"

Together with his friend Lincoln Schuster, Simon founded a publishing company called Simon & Schuster and released *The Cross Word Puzzle Book*, a collection of old puzzles from the *New York World* that came with a pencil attached. The book became a sensation. By 1925 more than 300,000 copies had been sold and Simon & Schuster was on its way to becoming one of the biggest book publishers in the United States.

After that the world went crossword crazy. Crosswords began appearing in newspaper after newspaper and people became obsessed with solving them. Some took their puzzle-solving way too seriously. In 1924 one Chicago woman filed for divorce because her husband stopped going to work so he could focus on his crosswords. The following year a man shot his wife because she refused to help him with a particularly vexing crossword.

By the time Butts turned to crosswords for inspiration, the initial excitement had died down, but these brainteasers had become a regular feature of newspapers across the world.

Drawing on the structure of crosswords, Butts divided his board into a grid of squares. Players would put their letter tiles onto the squares and, just as in crosswords, the words they added would have to fit in with the words already on the board.

Butts felt the game needed a better scoring system too. So he went back over his records of letter frequencies and began calculating how many points each letter tile should award players when used in a word. Common letters like A and E would earn a single point, mildly frequent ones like C and P would give three points, and the rarest letters of all, Q and Z, would boost a player's score by a massive ten points.

Finally, to give the game an element of strategy, Butts added special squares to the board that would double or triple the word or letter scores if a player placed a letter tile on it during his or her turn.

Butts spent months tweaking, refining, and polishing his new board game. He toyed with board sizes, experimented with the locations of double and triple scores, adjusted letter frequencies and points, moved the start position around, and added blank tiles that could be whatever letter a player wanted it to be.

When he finished the game it was 1938 and Butts's old architectural practice had rehired him. The years of scraping by were over. Nonetheless Butts still believed in the game he had spent years perfecting and decided to try selling it.

He initially sold his new game under the name "It" before changing the title to the more descriptive Criss-Cross Words. Butts depended completely on word of mouth to sell his game and as a result sales were dismal. When a rare order arrived, Butts would carefully make the letter tiles, draw the board, mimeograph the rules, and hunt for a suitable box. After doing this around hundred times, Butts gave up. No one seemed to want his game and making copies of it was just a chore with little reward.

The game's story could well have ended there. Years went by. The Second World War started and finished. Butts got on with his life. Now and again he would play Criss-Cross Words with his wife or his friends but most of the time his game sat on the shelf ignored and unloved.

Then in 1947 Butts got an unexpected call from a man named James Brunot. Brunot lived on a sheep farm in Newtown, Connecticut, but worked in New York City as a social worker. He wanted to start a small business so he could escape the draining four hours a day he spent commuting and spend more time with his wife.

Brunot had been introduced to Criss-Cross Words by a social worker who knew Butts, and after discovering the game wasn't in production he thought the game could be his ticket out of the rat race. So he called Butts and asked if he could buy the rights. Pleased that someone, at last, had shown an interest in his game, Butts sold him the rights to make the game in return for a royalty on sales.

After getting the go-ahead Brunot spruced up the look of the board, added a fifty-point bonus for players who put all seven of their letter tiles down in one go and made the starting square a double-word score. Finally he gave the game a new name: Scrabble.

In summer 1948 he hired local woodworkers to make letter tiles, ordered boards to be printed, and then began assembling each copy of Scrabble on his kitchen table.

Brunot might have had enough money to put the game into production but he had little left for promotion and Scrabble floundered. In 1949 he sold around 2,400 sets and the following year sales dropped. In 1951 he managed to push the sales figures up to almost five thousand copies but he was still losing money. Sales continued to pick up as the summer of 1952 approached but nowhere near fast enough to convince Brunot that Scrabble was worth sticking with. As Brunot and his wife headed to Kentucky for a week's vacation, he began thinking about shutting the business.

So it came as something of a shock when they returned from Kentucky to find an order for 2,500 copies of the game. The following week orders for a total of 3,000 Scrabble sets arrived. The next week the order total was even higher. What was going on?

Brunot never really figured it out. Maybe Scrabble's drip-drip of sales had achieved a critical mass and word of the game was rising exponentially. Another possibility was the rumor that Jack Straus, the chairman of Macy's, had played Scrabble on vacation and on returning to New York ordered his stores to stock the game. And once Macy's started stocking Scrabble, other retailers followed suit in case Macy's knew something they didn't.

The flood of orders overwhelmed the Brunots. Their home filled up with boxes, boards, racks, and letter tiles until they could barely move around the place. They moved the business to an old schoolhouse. It filled up too. They relocated the business again, but even with a team of thirty-five employees putting together six thousand sets a week, they still couldn't keep up with demand.

Unable to cope Brunot licensed the production rights to the game manufacturer Selchow & Righter in 1953 and concerned himself with making a deluxe edition that cost ten dollars and came in a red imitation-leather case.

Even Selchow & Righter struggled to meet demand for Scrabble. Whenever the game appeared in stores, shoppers quickly cleared the shelves. By the end of 1954 close to four million Scrabble sets had been sold, and while demand eased off in the following years, annual sales rarely dipped below a million copies for the rest of the decade.

With Scrabble taking the nation by storm Brunot found himself besieged with questions from players about what words were permissible in the game. Was pa ok? What about whoosh? Is aspirin allowed or does it count as a trademark? "I don't give a damn, use whatever words you want to," was Brunot's bewildered take on it all.

Brunot was equally mystified by the dictionary publishers begging for the right to be declared the official Scrabble dictionary. Merriam-Webster even sent him a draft Scrabble dictionary, a giant paper list of words pulled from their standard dictionary but with their definitions missing. What nonsense is this? Brunot probably thought.

He did approve the release of *The Scrabble Word Guide,* a definition-free list of thirty thousand words that its two authors thought Scrabble fans might find useful. But Brunot drew the line at the idea of an official Scrabble dictionary and turned down every offer the publishing world made. "It's only a game," he told *Life* magazine. "It's something you're supposed to enjoy."

How times change. Today the English-speaking world has not one but two official Scrabble dictionaries to choose from.

The player passion that flummoxed Brunot never went away; in fact it intensified. It was as if the obsessiveness Butts applied to making Scrabble had seeped out of the board and into the minds of players.

The arguments about what words were and were not acceptable in Scrabble created a world filled with armchair linguists. They would clash over whether a word was a proper or common noun, discuss whether ad was an abbreviation or a word in its own right, develop theories to prove that "back yard" was really "backyard," and got into wrangles about the use of long-forgotten words like "gardyloo," the warning that people in Edinburgh cried when emptying chamber pots out of windows so that those in the streets below could take cover.

In the United States players wondered if British spellings like "colour" were acceptable while players across the Atlantic pondered whether the rule barring foreign words applied to American spellings.

Obscure words containing high-scoring letters became prized discoveries among those who took their Scrabble seriously. Not only could such words boost their score and rid them of troublesome tiles, they also got to feel smug when their opponents challenged it only to find it in the dictionary.

The limits of Scrabble were being tested and despite Butts's perfectionism the game's rules simply couldn't deal with every curveball thrown by English, a bastard language stained with the legacies of centuries of invasion and conquest. Consult a dictionary was the best solution the rules could offer but that only led to arguments about the merits of different dictionaries.

The task of trying to resolve these debates only really began when Brunot and Scrabble parted ways. In 1971 he sold the North American rights to Scrabble to Selchow & Righter for more than a million dollars, by which time he had also offloaded the rest of the rights to the London-based company Spear's Games.

Butts did well from these deals too. While his royalties where never enough to allow him to buy a millionaire's mansion, it was more than enough to enable him and his wife to stay in Jackson Heights and enjoy the rest of their lives in comfort.

After purchasing the rights, Selchow & Righter, possibly inspired by the Bobby Fischer versus Boris Spassky world chess championship match of 1972, decided to try to keep Scrabble in the public eye by holding tournaments around the United States. The goal was to turn the popular board game into a competitive mind sport and it worked.

After low-key trials in New Jersey and Pennsylvania, the first official Scrabble tournament was held in Brooklyn during March 1973 with support from New York City's Department of Recreation. A few months later Selchow & Righter began taking Scrabble tournaments nationwide. Meanwhile, Spear's Games had hit on the same idea and started organizing Scrabble contests for British fans of the game. The tournaments on both sides of the Atlantic were a success, generating publicity and fostering a community of dedicated players who soon took over the task of organizing the competitions.

But competitions need rules and it soon became clear that some definitive answers to all those vocabulary questions would be needed. So, as the game's rules advised, the owners of Scrabble turned to dictionaries.

Selchow & Righter gave Merriam-Webster the deal it had waited years for, and in 1978 its *Official Scrabble Players Dictionary* arrived in North American bookstores. A decade later, after years of muddling through with standard dictionaries, the rest of the world finally got an official word list based on *The Chambers Dictionary*.

Players now had dictionaries tailor-made for Scrabble. No foreign words, no proper nouns, and no distracting definitions; just words and their scores, words reduced to a numerical essence. Gardyloo was no longer a word that revealed something about life before the installation of sewers; it was thirteen points.

And when words have no meaning beyond the points they deliver, you no longer had to even speak the language to be a Scrabble champion. In July 2015 Nigel Richards, a New Zealander with a bowl haircut and beard worthy of Texan rockers ZZ Top, took part in the French-language Scrabble world championship in Belgium. Despite barely being able to order lunch in French, he won. Richards, who has also won the English world title three times, prepared for the contest by spending nine weeks memorizing the words in the official French-language Scrabble dictionary.

Richards is living proof that despite its appearance Scrabble is not a game of words but a game of memory and numbers. "A lot of these guys are computing geeks and so on, so they are quite into programming and they've got that sort of logical mind," says Gerry Breslin, publishing manager at Collins, the publisher that replaced Chambers as the producers of the *Official Scrabble Words* dictionary for players outside North America in 2003. "For them it's really all about the score and about strategy; how to close out the game and block up an opponent."

Given this angle, putting together a Scrabble dictionary requires looking at words in an unusual way, says Mary O'Neill, project managing editor at Collins. "Usually we are looking at words from the point of view of how we describe their meaning, and when you're working with Scrabble you have got to look at things in a whole new way," she says. "You're looking at words from the point of view of playability. When you're doing a general dictionary you're not thinking about whether it's a two-letter word and how many points you have and how many Y's and J's there are."

Since words are point-scoring devices, top-tier players are always lobbying for new words to be added to the dictionaries. Collins, for example, regularly gets sent lists of Maori words from a player in New Zealand who hopes to get an edge in the game by having them included in the dictionary. "Sometimes they try to force the pace a little bit and we have to just say there is a due process for these words to make it into the dictionary," says Breslin.

The process of getting new words into the Scrabble dictionaries reflects the global split in the ownership of the game. Today the North American rights to Scrabble are owned by Hasbro while Mattel controls the game everywhere else in the world.

For Collins, Mattel's chosen lexicographers, compiling a new Scrabble word list is a four-year cycle that starts once the latest edition of the *Collins English Dictionary* is ready. Once a word gains enough legitimacy to be added to the main dictionary, Collins will then consider it for *Official Scrabble Words,* provided it complies with the rules of the game.

The task of assessing what new entries should be candidates for the Scrabble word list falls to the dictionary committee of the World English-Language Scrabble Players Association. The committee painstakingly goes through each new word, picking out possible options and throwing in some suggestions of its own. The committee, a volunteer team of expert Scrabble players, and Collins then debate and finally agree on what words will be added to the next edition of *Official Scrabble Words*. Sometimes words suggested by the players even make it into the main dictionary, such as "bemix," which means to mix thoroughly and first appeared in the Scrabble word list. Beyond ensuring Collins adheres to its branding guidelines, Mattel stays out of the process.

In North America, however, players take a backseat and Merriam-Webster decides what words from its main dictionary will go into the *Official Scrabble Players Dictionary,* which the company has been publishing for almost forty years. "We haven't had much input from Scrabble players about adding words," says Jim Lowe, general editor of the *Official Scrabble Players Dictionary.* "They just accept the words that we put into the dictionary. They know they are constrained by that at tournaments."

While the players are less involved, Hasbro takes a more hands-on approach to its official dictionary than Mattel and has ruled that profanities and words that could cause offense must not be included.

The cleaning up of the North American Scrabble dictionary began in the 1990s after a Holocaust survivor discovered that her official Scrabble dictionary contained the word "kike" among other offensive words. After complaints from her and the Anti-Defamation League, Hasbro ordered a cull that saw around 175 problematic words removed from the official Scrabble dictionary, including poo, redneck, jigaboo, bollocks, and arse. Offensive words with double meanings, such as faggot or bitch, survived.

"The reason for removing offensive words is that the publisher Hasbro is promoting our dictionaries in schools and for families," says Lowe. Not that filtering out what is and is not offensive is clear-cut. "A lot of it has to do with political correctness. We have to look at the context and see how these words are used and whether they are meant to be disparaging or derogatory or just plain vulgar. If they are we won't put them in the dictionary."

Collins, in contrast, leaves the profanities in place. Helen Newstead, head of content at Collins, says this stance not only pleases tournament-level players who want as many words at their disposal as possible but is also in keeping with wider trends in dictionaries. "Since the 1950s the feeling among lexicographers has been that, as lexicography is descriptive rather than prescriptive, and as many people do use offensive language, dictionaries for general adult use should include offensive words," she says.

But even once the removal of offensive words is factored in there is a huge gulf between the number of words in the lists used for North American tournaments compared to the rest of the English-speaking world. While Collins gives the rest of the world around 260,000 words to play with, the official list for North America contains just below 190,000 even after profanities are put back in.

This has been a bone of contention for North America's top players for years. Expert players crave more words since every additional one opens up more possibilities in the game, and when the world's best players come together for the World Scrabble Championship every two years, North American players have to learn the thousands of extra words on offer from Collins and then somehow forget them when they return to competitions in the United States and Canada.

As a result the North American Scrabble Players Association has, despite relying on Hasbro for funding, permitted some U.S. tournaments to use the Collins word list. At the moment just two hundred of the fifteen hundred or so tournament-level players in the United States use Collins, but the number is growing.

The main reason why the Collins list has so many extra words is because *Official Scrabble Words* reflects English in all its forms. American, British, Irish, Australian, South African, and Indian English words are all in there. Merriam-Webster, however, limits itself to U.S. and Canadian English.

Ultimately the gulf between the world's two official Scrabble dictionaries is more a by-product of the diversity of English itself, a diversity that is always increasing as new words appear. "There's no end to these new words," says Lowe. "They just keep coming in and coming in. All we can do is try to keep track of them and add them into the dictionary."

One change Lowe expects to see in years to come is the integration of more Spanish words into English thanks to the influence of the growing U.S. Hispanic population. That integration into dictionaries will, however, take time. "It's the words that have been borrowed and taken into our language that we're interested in, not just Spanish words themselves. They have to be used by American writers," he says.

Not that any of that matters to the world's best Scrabble players of course. After all for them it's all about the points to be scored.

10

# PLASTIC FANTASTIC: MOUSE TRAP, OPERATION, AND THE WILLY WONKA OF TOYS

## *How Mouse Trap and Operation took board games into the plastic age*

When Marvin Glass called, the toy and game industry jumped.

By the middle of the 1960s everyone in the toy world wanted to know what this five foot three, chain-smoking ball of nervous energy would do next. At trade shows people would clamor to see what he was up to and in their offices corporate executives daydreamed about receiving an invitation to Chicago to see the latest wonder to emerge from his secretive toy workshop. What mad, yet marvelous, idea would Glass come up with next?

The mind of the wiry man at the center of their attentions had always fizzed with visions of off-the-wall playthings. The son of Jewish immigrants from Germany, Glass was born on July 14, 1914, in the Chicago suburb of Evanston.

One of his earliest memories was exploding with rage as a four-year-old because the cardboard dog he had made wouldn't wag its own tail. His dreams of better toys continued throughout his childhood. By eight he was building submarines that fired wooden torpedoes and making Roman helmets, swords, and shields for him and his friends. "I always played Caesar and I never got assassinated," he told the *Saturday Evening Post*.

Glass's time spent prancing around the streets dressed as a Roman emperor did little, however, to ease the boy's loneliness. His relationship with his unhappily married parents was fraught at the best of times. His father, a six-foot-tall engineering consultant, would wonder aloud about his son's meager frame and found the boy's daydreaming a source of endless frustration.

"What do you want to be when you grow up?" his father would demand. "Nobody," Glass would snarl back to his father's fury.

Eventually his parents packed him off to a private military school in Wisconsin, doubtless hoping it would straighten their son out. Glass hated the school and retreated more and more into the world of toys, constructing models of pirate ships and Egyptian galleys to distract him from his misery. Toys, he later told reporters, were an escape from a "sordid and unsatisfactory world." Toys cars never had fatal accidents, tin soldiers were never blown apart on the battlefield, and baby dolls never grew old. Toys were the world cut loose from misery, pain, and the specter of death.

The military school failed to change Glass and neither of his parents attended his graduation. After completing his psychology degree in 1935, Glass moved into a two-bedroom apartment on Ohio Street, in Chicago, with his friend Eoina Nudelman, and it was there that the boy who wanted to be nobody became somebody.

Nudelman designed store window displays and one day a customer asked if he could come up with an idea for a toy. He asked Glass for help and together the pair devised a projector that kids could use to illuminate comic strips. They sold their creation for five hundred dollars, which seemed fantastic until they learned that the manufacturer went on to earn more than thirty thousand dollars from the invention. Glass vowed then and there never to sell another idea without securing a royalty as part of the deal.

Flush from their early success, the friends set themselves up as toy makers and began conjuring up fun ideas on a diet of vodka, cigarettes, and Romany music. Eventually they hit pay dirt with a series of paper dolls based on famous cartoon characters like Mickey Mouse.

In 1941 Glass ended the partnership and struck out on his own. He borrowed eighty dollars, formed Marvin Glass and Associates, and moved into a loft on Rush Street. A few years later he created the unlikely product that would make him his first million: the Catholic Weather Chapel, a cuckoo clock-like structure housing a moisture gauge. When it was dry a Sacred Heart would pop out of the chapel, but when the moisture levels suggested bad weather, Saint Barbara, the protectress against sudden death from natural disasters, would emerge. It was a bizarre and tacky creation but it sold really well.

Unfortunately, Glass's next creation lost him his first million. He got the idea one night when he passed an antiques store on Michigan Avenue that had a stained-glass window on display. On seeing the antique his imagination whirred into action and he soon convinced himself that stained-glass Christmas tree decorations made from plastic would be enormously popular. He splurged more than a million dollars on making and promoting his mail-order trinkets and recruited a team of office girls to process the orders and a security guard to make sure none of them stole the money customers sent in. Then he sat back and waited for the orders to arrive in the mail.

The deluge he expected never came and the few people who did place orders soon began demanding their money back due to the dismal quality of the decorations. After that Glass vowed that he would never again manufacture his inventions. Instead, he would sell his creations to manufacturers and they could carry the risk.

Whether Glass would last long enough to make another product was uncertain, however. His Christmas tree misadventure had left him saddled with huge debts that threatened to engulf him and his business. He needed a hit, and fast.

In a stroke of good fortune a young man named Eddy Goldfarb knocked on his door. Goldfarb was also a toy inventor and he wanted to show Glass a cardboard chicken he had created that laid marble eggs when the box that formed its body was pushed down. Glass loved it and hired Goldfarb. They refashioned the cardboard hen in red Bakelite plastic and sold it to a Chicago toy manufacturer who released it as the Busy Biddee Chicken in 1948.

By the end of the year more than ten million Busy Biddee Chickens had been sold, rescuing Glass from his self-inflicted mountain of debt. Goldfarb then came up with an even better idea: a set of wind-up chattering dentures called the Yakity-Yak Talking Teeth. These plastic teeth sold even better than the Busy Biddee Chicken.

After that not even Goldfarb's decision to move to California and start his own toy invention agency could hold Glass back. The 1950s saw Glass and the toy inventors he hired delivering smash-hit toy after smash-hit toy. The inventions that poured out of his Chicago workshop imprinted themselves in the childhood memories of a generation.

There was Moody Mutt, the multimillion-selling toy dachshund that would bear its fangs if pushed from behind; Merry Go Sips, a million-selling drinking cup where cartoon animals spun around inside its transparent lid when toddlers sucked their milk through its straw; and Ric-O-Shay, a toy pistol that Glass and his team worked on for two years to ensure that it sounded exactly like a gun from a Western movie.

Another runaway success was fake barf, the icky brainchild of employee Carl Ayala. Ayala got the idea after encountering somebody's vomit on a sidewalk. At first he recoiled in disgust but then inspiration struck. He raced to work and over the next few days fashioned a puddle of sick formed out of yellow latex and foam chunks. After finishing his work he showed his prototype to Glass, who promptly dismissed it as an utterly vile creation.

Sometime later Glass invited Irving Fishlove, the owner of a Chicago-based novelty manufacturer, to the workshop for a showcase of the company's latest ideas. Things went badly. Fishlove sat unmoved by everything Glass put before him. But then Ayala burst into the room and slapped his latex barf on the table. Fishlove burst out laughing and Glass's showman instincts kicked in. Of course we saved the best for last, Glass declared as if Ayala's interruption was always part of the sales pitch. By the end of the meeting Fishlove had agreed to buy the rights to Ayala's product. When the fake barf launched under the name Whoops, it rapidly joined Marvin Glass and Associates' roster of successful inventions as the nation's practical jokers jumped at the chance to gross out unsuspecting victims.

Not everything the company did worked. Among the disappointments was Crybaby, a doll that would cry if its pacifier was removed. Glass blamed its failure on people seeing it as sadistic rather than funny. But with so many of his company's creations striking gold, such failures did little to dent the hit factory's reputation.

The company's successes not only made Glass rich but changed the way the toy industry operated. Before Marvin Glass and Associates, toy design was usually done in-house by manufacturers who were dismissive of anything invented outside their own walls. But Glass's winning streak forced them to think again. Glass's creations were so different, so marvelous, and so popular with the public that they simply could not be ignored.

These wondrous designs, coupled with Glass's almost hypnotic showmanship, dazzled toy manufacturers and in the process got them to open their doors and their minds to the countless independent toy agencies that would ape Glass's example.

By the mid-1950s people were calling Glass the Walt Disney of toys, but with his air of sadness, nervous mannerisms, and tendency to flip from charmer to tyrant in an instant, today Roald Dahl's eccentric chocolatier Willy Wonka seems a more apt comparison.

And with each new hit, the parallels with Wonka only deepened. The paranoia that prompted Glass to hire a guard to watch his employees as they processed orders for his Christmas decorations also led him to turn his workshop into a top-secret toy lab. Glass fretted constantly about the not entirely outlandish risk of corporate espionage and responded by transforming his offices within the Alexandria Hotel in Chicago's North Side into a clandestine warren of rooms that were off-limits to outsiders. The doors separating each area were fitted with multiple locks that were changed regularly. Trashcans were secured with padlocks to prevent pilfering and employees were sternly warned that it was forbidden for them to discuss their work with any outsiders, including their husbands or wives. Glass even wished there was a way he could stop his designers from talking in their sleep.

Visitors to the premises, meanwhile, would find themselves confronted with a locked bright red entrance door with a peephole through which a distrusting eyeball would scrutinize them before deciding whether to let them into the reception area. One employee said it was like working in an underground bunker.

Glass's workaholic lifestyle and his expectation that his employees match his own frenetic approach to working life also made him a difficult man to work for. Phone calls from Glass in the dead of the night were commonplace, as were demands for weekend meetings.

Glass also had zero tolerance of any ideas he regarded as substandard. "There was only A. There was no A-minus or B-plus," says Jeffrey Breslow, who joined the company in 1967 and later became its president. "If it was A-minus or B-plus it was shit. There was no settling for anything else. In the early days Marvin would throw prototypes across the room."

Even the struggles of Glass's failed marriages became fuel for his creativity. One night Glass ended up having a blazing row on the phone with his ex-wife, who accused him of having become nothing more than a machine, a cold inhuman automaton. She meant to insult him, but Glass found her barbed words inspiring. The result was Mr. Machine, a wind-up robot with a stovepipe hat and a transparent plastic body that put the brightly colored gears of its clockwork heart on display for all to see. Once wound up, Mr. Machine would march forward, arms and legs swinging, as the alarm bell in his belly rang out and his mouth opened to emit an almost pained squawk. Mr. Machine could also be disassembled and reassembled at will by the children that controlled his fate.

"Mr. Machine is modern man tyrannized by his mechanical creations, turned into a mechanical man himself," Glass told *Life* magazine about the creation that he regarded as himself rendered in toy form.

Mr. Machine became another Christmas list topper. Glass sold the idea to the Ideal Toy Corporation and on its release in 1960 it instantly became one of the New York toy firm's best sellers. Mr. Machine proved so popular that the following year Ideal Toy released the Mr. Machine Game, a board game where players raced little Mr. Machines back to the factory across a network of winding and crisscrossing paths.

While Marvin Glass and Associates didn't develop the Mr. Machine Game, the company, which had spent the 1950s shaking up the toy business, was now looking to do the same for board games.

One of the company's first steps into the world of games came shortly after Mr. Machine in the form of 1961's Miss Popularity, a now uncomfortably sexist game where teenage girls compete to become the most admired largely on the basis of their looks.

Central to the game's appeal was the inclusion of a pocket-size mechanical phone formed out of lavender-colored plastic. As the game progressed players collected cards that could help or hinder their efforts to become popular. They might learn that their pretty legs landed them a rewarding job as a hosiery model or find they lost a hefty chunk of popularity points because they had "neglected their personal appearance." Other cards instructed players to dial a number on the phone to find out the answers to pressing yes-no questions such as whether a beauty parade judge thinks their measurements mean they have a nice figure. On dialing the phone, it would whir and up would pop one of four random answers: yes, no, maybe, or busy.

Miss Popularity's view that teenage girls should tailor their entire lives for the pleasure of creepy adult men speaks volumes about attitudes in the early 1960s, but the game's fusion of toy phone with static cardboard provided an early sign of how Glass and his ideas factory intended to bring fresh thinking to board games. And in 1962 Marvin Glass and Associates showed just how far their plastics-enhanced vision for board games could go when Ideal Toy published its game Haunted House.

Sold in a vast box that was almost twice the size of a standard board game, Haunted House whisked games into the plastic age. Instead of a flat board, Haunted House had a three-dimensional rendering of an *Addams Family*–style mansion that stood upright during play. The squares of traditional games were gone, replaced by round holes into which players would place their pegs as they roamed the various rooms of the spooky home.

Nor were there dice. Instead there was a mechanical "owl spinner" that responded to pulls of its lever with the sound of hoots before displaying how many spaces players could move. The bells and whistles didn't stop there. The board was peppered with mechanical features that caused panels to spring open to reveal vampires and ghosts and the treasure players were searching for.

Behind its attention-grabbing plastic facade Haunted House was just another race game where the outcome was left to chance, but it was perfect for television. "Kids were watching TV and that was the medium to get your product to the audience more than anything else so, basically, Marvin Glass and other companies at that time were making television toys," says Breslow. "If it wasn't visual, if it didn't move, jump up and down, and do stuff, it didn't tell a story. It's very hard to do TV commercials with board games. It's hard to make a TV commercial about Monopoly."

Despite its televisual appeal, Haunted House still wasn't the success Glass had hoped for, but the following year his company delivered a board game that could join Mr. Machine and the Yakity-Yak Talking Teeth on the bestseller list.

The game started with Glass and Burt Meyer, one of the company's partners, sitting around trying to come up with a fresh idea. "We were in Marvin's office and looking here and looking there, and he picked up a newspaper and we started thumbing through that," remembers Meyer. A few pages into the newspaper they found a cartoon called "How to Remove the Cotton Out of a Bottle of Aspirin" by Rube Goldberg. The cartoon was typical of Goldberg's work: a humorous and ludicrously overdesigned contraption that solved a simple everyday task.

The outlandish process Goldberg envisaged caught the pair's imagination. "Maybe we could do something with that?" said Glass. A game maybe, agreed Meyer. They began thinking about what kind of Goldberg-style apparatus the game should have. Removing cotton from an aspirin bottle wouldn't work; after all it was hardly the kind of task to fire children's imaginations. Meyer suggested a mousetrap.

Armed with the idea for a board game featuring an over-the-top mousetrap, Meyer left the office and asked designer Gordon Barlow to make it happen. "We didn't really drive people to do exactly what we said, so I went in a couple of days later to his desk and said: 'How's the mousetrap coming?'" says Meyer. "He said: 'Oh, I didn't like that, I'm not going to work on that one.' I said: 'Gordon, that's really going to be a good idea, it'll be worth our while and if you stick with it. I'll help you.'"

So the pair began plotting out the game. They imagined a ridiculously convoluted trap with sixteen parts. First a crank would be turned to rotate a set of gears that would cause a lever to move a roadside stop sign so that it hit a shoe on a stick. The shoe on the stick would then kick a bucket causing the metal ball inside it to roll down a rickety staircase and a drainpipe. At the end of the drainpipe the ball would nudge a vertical rod with a hand on top, which would push a bowling bowl off a ledge and into a bath. The bowling ball would then drop through a hole in the bathtub and land on a seesaw. As the ball upended the seesaw a man in a bathing suit at the other end would be propelled into the air backward and land in an old-fashioned washtub. The force of the man landing in the washtub would then unbalance a cage perched on top of a post so that it dropped down and—finally—captured the mouse below.

Contraption in place they then created a simple race game where players moved mice around the board based on dice rolls. As they moved toward the cheese at the end of the path they would land on squares that enabled each part of the mousetrap to be put into place. Once the mice reached the far end of the board they would circle round and round, until all but one of them had been caught by the trap.

As a game it was rather dull but it didn't matter because how it was played was secondary to the thrill of seeing a Goldberg-style invention brought to life in colorful plastic.

Pleased with the result, Glass and Meyer took the prototype to Milton Bradley's headquarters in Massachusetts in the hope of securing a deal. James Shea Sr., the president of Milton Bradley, hated it. "This is not a game, this is nothing," he told the pair. "A game, you play it on a board and you roll dice but this, this is a lot of plastic junk. We can't use this."

Even Glass's magnetic personality couldn't change Shea's mind. So they took the game to Milton Bradley's archrivals Parker Brothers, only to leave disappointed again. "Parker Brothers essentially said the same thing in nicer terms," says Meyer.

Having been snubbed by the nation's leading game publishers, Glass decided to see if Ideal Toy could be tempted by Mouse Trap. Ideal Toy's president Lionel Weintraub was skeptical. Sure they had released Haunted House but that looked more like a toy than a game and Mouse Trap was most definitely a board game. "I'm not in the game business," he said after they showed him the prototype. "Well, this will put you in the game business," replied Meyer. Weintraub thought for a while and then said: "Ok . . . I'll give it a try."

Few in the toy and games business rated Mouse Trap's chances. "There was a great deal of skepticism from the trade," says former Ideal Toy employee Philip E. Orbanes. "Who's going to pay all this money for this plastic toy layered to a game board? And besides it's not very exciting—you go around the board until the end and your mouse goes around until it's trapped."

Unsure whether the game was genius or folly, Ideal Toy started by launching Mouse Trap in Pittsburgh, Pennsylvania, with some local TV advertising to see how it fared. "The result was they couldn't make enough," says Orbanes. "They sold something like three million copies in the next twelve months."

Meyer's claim that Mouse Trap would put Ideal Toy in the games business was spot on. In a matter of months Ideal Toy went from a toy company that had dabbled in games to the third biggest game publisher in the United States. "When Ideal Toy put it on the market and both Milton Bradley and Parker Brothers saw the reaction to it, which was tremendous, they were a little perturbed that they passed on it," says Meyer. "Especially Milton Bradley. Parker Brothers was more pristine as the leader of the board game industry. Milton Bradley just saw that they had missed an opportunity. Ideal Toy loved it because it put them in the plastic action game business."

As Glass told the *Lawrence* (Kansas) *Journal-World* newspaper in 1965: "It's knowledgeable people who hold back progress in the toy industry. They have all the answers, and the answers are always wrong."

After Mouse Trap, manufacturers flooded the store shelves with plastic action games that blurred the divide between toys and board games. Some of their games became enduring classics, among them 1965's Trouble, a simplified take on Parcheesi that sold largely because of its Pop-O-Matic die roller—a plastic dome in the middle of board that spun the die when pressed—and 1974's Connect Four, where players vied to be first to get four counters in a line on its vertical plastic board.

Marvin Glass and Associates also produced plenty more plastic action games, including 1965's Fish Bait, a fisherman-themed take on Mouse Trap, and 1966's Babysitter Game, where players had to roam the board completing household chores without waking the ugly single-toothed mechanical baby asleep in the cot at the middle of its plastic board. The most successful Marvin Glass game to follow in Mouse Trap's footsteps, however, was, by some margin, 1965's Operation.

The game didn't begin life within Glass's secretive workshop but at the University of Illinois, where industrial design student John Spinello was given an assignment requiring him to devise an electric game. He responded by building a metal box into the top of which he drilled a series of holes and cut a meandering groove. Players had to insert a metal probe into the holes and carefully guide it along the groove he had cut. If the probe touched the sides it would connect the circuit and set off a loud buzzer hidden within the box. His impressed tutors gave him an A for his work.

Spinello showed his battery-powered game to his godfather Sam Cottone, a model maker at Marvin Glass and Associates. Cottone thought it was great and arranged for Spinello to show it to Glass. But Glass was far from impressed when Spinello placed his metal box on the desk in front of him. What is this trash? demanded Glass. It was not a good start, but Spinello pressed on. He explained how to play it and then handed the probe to Glass.

Glass began disinterestedly guiding the probe through the higgledy-piggledy groove in the box and then it touched the side. The buzzer screamed into life and a spark jumped between the probe and the box. Glass threw the probe into the air, crying: "I love it! I love it!"

Glass offered Spinello five hundred dollars for the rights and promised to give him a job after he graduated. Spinello agreed. Five hundred dollars was a hefty chunk of money in 1964, enough to cover a whole semester of university tuition, and having a fun job waiting for him when he finished his studies would be ace.

The job never happened. For a few months Glass made excuses about the office not being ready before getting his lawyer to call Spinello and tell him that there would be no job after all.

In the meantime Glass's inventors had been busy fashioning Spinello's creation into a marketable game. At Cottone's suggestion they made it a game about searching for water sources in the desert. The box became a desert landscape filled with holes and cracks to probe and the company began calling it Death Valley.

The prototype caught the attention of Milton Bradley executive Mel Taft, who bought the rights and then showed the game to Jim O'Connor, one of the company's in-house game designers. It would be better if players had to remove objects from the holes without touching the sides using tweezers, O'Connor suggested.

So Milton Bradley reworked the design and turned it into a tongue-in-cheek surgery game where players had to pull spare ribs, rubber bands, stomach butterflies, and other oddities out of the innards of patient Cavity Sam without setting off the buzzer and lighting up his great big red nose.

Like Mouse Trap, Operation was one of those rare hit games that instead of burning out and fading away became a permanent feature on game-store shelves. Operation went on to earn tens of millions of dollars over the decades that followed its debut.

Spinello could only watch from the sidelines as the raw idea he presented raked in the big money. "It's kind of a sad story," says Breslow. "But who knew it was going to become Operation? Certainly not Marvin, certainly not John. The nature of the toy business is that ninety-five percent of what we do never sees the light of day."

For Glass, however, his five hundred–dollar investment paid off handsomely and by the early 1970s an estimated one in every twenty toys and games being sold in the United States began life in his Chicago toy lab.

Yet all the money and acclaim did little to ease the sadness and streak of paranoia that lurked within the man behind the thrills of millions of childhoods. He told reporters that he regarded himself as a "complete and utter failure" and that the amount of money he earned was obscene.

Glass's dynamo lifestyle showed no sign of slowing down either. He continued to sustain himself each day by puffing his way through three packs of cigarettes and a dozen cigars, drinking countless cups of coffee, and nibbling a sandwich or two. And at night he would grab no more than four or five hours of sleep. "I dread sleeping longer," he told the *Saturday Evening Post*. "It's like being dead."

Instead he threw glamorous parties at his home in Evanston, a large hundred-year-old carriage house he transformed into a luxurious pad decorated with Picasso and Dalí paintings and Frank Gallo sculptures. But even as his guests partied in the vast Jacuzzi whirlpool and danced to the music played by show-business stars on his grand piano, work was never far from Glass's mind. He would rope guests into trying out the latest creations to emerge from his toy lab and even had a phone installed in the basement sauna so even amid naked frolics in the hot tub a direct line to work was always in easy reach.

As well as lavishing money on his own home, Glass splashed out on a custom-built headquarters for his company on the corner of La Salle Street and Chicago Avenue. But while the money turned his home into a rival for Hugh Hefner's Playboy Mansion, it turned his toy lab into an imposing fortress.

An almost Brutalist slab of windowless concrete with two-foot-thick walls, the two-story building had close circuit television cameras scanning the area around it and an iron gate manned by a security guard. Inside napkins were banished from the canteen lest a designer doodled something of significance on them, and every night the inventors had to gather up their work and take it to be locked inside a giant bank vault buried deep within the building. Any designer who forgot to put their work in the vault before heading home would face a severe dressing down the next morning from an enraged Glass in full-on Caesar mode.

The facility also housed art and sound studios, a chemical lab, and even a miniature factory for the production of prototypes. And right in the middle of the building was Glass's office, which was protected with double walls in case anyone was trying to listen in remotely.

Glass even began arriving at trade shows in an armored truck with armed guards in tow and the suitcase containing his latest prototypes handcuffed to his wrist. Where the lines between paranoia, publicity seeking, and protecting his inventions now were was impossible for anyone to tell.

Not that Glass treated his inventors as inmates at his toy lab prison. Far from it. The building's entrance hall was decorated with expensive nude bronze statues and he would hire expert chefs to cook lunches for the team. And while his demands and domineering side sometimes left employees on the verge of tears, he also inspired loyalty with his charm, generous bonus payments, and support when close relatives passed away.

Yet Glass remained a troubled man who worried that every headache was a symptom of a brain tumor but refused to take pills in case they made him stupid.

"I would characterize Marvin as a paranoid schizophrenic

personality—not that I'm a psychiatrist," says Breslow. "You went to work there and he says you can't talk to your wife about anything, you can't talk to anyone outside this building about anything we're doing inside. 'I find out that you're talking about stuff, you're fired.' He had a secretary for many years and he never let her in the back of the studio, and she worked with him and she was his friend and he wouldn't let her behind the doors. He was a very complicated man."

Glass was also utterly terrified of flying and never more so than when he was told that the British toy industry wanted to name him toy man of the year at the country's 1970 trade show in Brighton. Glass was over the moon about it until it dawned on him that collecting the accolade would involve a transatlantic flight.

"I got invited to go with him, so I was pretty excited," says Breslow. "So I go to the airport and he had his psychiatrist there to go on the plane with him. So we're at O'Hare Airport waiting to get on the plane and he's actually perspiring and he is white as a ghost—that's how worked up he was about getting on an airplane. Everybody told me be careful when you fly with Marvin, he's a nervous wreck, he'll make you crazy, and I didn't believe how bad it was.

"At any rate he was sitting behind me and his psychiatrist was sitting next to him and every time I got up he was giving Marvin pills to calm him down. I don't know what he was popping in his mouth but when we landed in London they needed a stretcher to carry him off the plane. He was totally knocked out. He didn't speak for a couple of days. We went a few days early and I think this was part of the reason."

But when a flight was make or break, Glass would pull out the stops to get on it, says Meyer: "One time, he was going to Cleveland to be interviewed on a radio talk show that was nationwide. We were driven to the airport and we were running late. We got to the gate and the jetway was disconnected and the airplane was moving back.

"We said, 'We've got to get on there,' and they said, 'Well, it's too late, you can't get on there.' And Marvin was a strong enough personality to say, 'We have to, we have to, we've gotta get on there, we have to go to a radio show in Cleveland, we're going to be on the air!' And with his manner, they called the airplane back. They put on the jetway, they removed two passengers because the airplane was full, and put us on. That typifies the personality he had, the power of persuasion he had."

Success might have allowed Glass to turn his toy lab into a bunker but his ambition remained intact, and in 1970 he railed against how consumers' unwillingness to pay more was hindering the evolution of games. "The next step for games will be a synthesis of the board and action-game types. There will be more family involvement, and games will find their way into more recreation rooms," he declared in *Toy and Novelties* magazine.

But, he added, thanks to people's reluctance to embrace higher prices, games were trapped in the technology of the 1940s and 1950s. "I'd like to see games in the same area as high-priced dolls, with greater play value and more involvement," he says. "When this happens we'll be able to put in new technologies, electrical devices, and so on."

Glass didn't live long enough to put his vision for the future of board games into action. The years of chain smoking and unhealthy living were taking their toll and in July 1973 he had a stroke. He was diagnosed with cancer soon after, and in January 1974 after months of illness, the Willy Wonka of toys and games passed away. He was fifty-nine.

"He was a unique character," says Meyer. "Some loved him, some people hated him. He did a little bit of creative thinking and designing, but his real forte was making creative people work. In other words, getting the most out of the people who worked for him. He was really creative in his handling of people and of designers, and that takes a talent. He just knew how to encourage them and when to berate them. He just knew how to touch people in the right manner verbally, sometimes encouraging, sometimes abusive. He just knew when to do what."

After Glass's death the company eased back on the security. Designers no longer had to seal their work within the vault every night and the all-hours working culture reverted to a more normal schedule. The changes didn't hurt and the hits continued to flow even though the company's charismatic founder was now a memory.

By the early 1980s the company had added electronic games to its list of smash hits including Simon, the iconic disc-shaped toy that challenged players to repeat the patterns lit up on its primary-colored surface, and Tapper, the popular 1983 coin-operated video game where players served beer to demanding patrons as fast as they could.

Marvin Glass and Associates' run of hits only came to an end in 1988, when the partners decided to go their separate ways and dissolve the business. The era of Marvin Glass was over. Even the fortress that the company once called home is no more, pulled down to make way for a new development.

Glass's company may be no more but the plastic-enhanced games his toy lab helped to pioneer live on. They may have lost their visual thunder to video games, but Operation, Mouse Trap, and Trouble are still on sale today and sit alongside more recent plastic action-game successes like Loopin' Louie and Pie Face.

Whether the lack of cardboard in these games excludes them from being considered board games is really a case of splitting hairs. "The terminology hasn't kept up with the time," says toy historian Tim Walsh. "We call them board games but there are plenty of games that don't have a board but people know what you're talking about. Operation is a game even though people might consider it a toy and it doesn't have a traditional board—it has a 3D board that is electrified. And when you look at Mouse Trap and Simon and a lot of the toys and games that came out of Marvin Glass and Associates, they are categorized as games. Whether they had a board or not doesn't really matter."

Glass's rewriting of the rules of the toy industry also endures. "Marvin Glass and Associates changed the toy industry by making the industry really recognize the independent inventor," says Meyer. "Throughout the last fifty years the industry has really depended on the independent inventor bringing things in and that's because of Marvin.

"It's also actually because of Marvin's ego because for a long time we were very secretive about what went on and we didn't tell anyone anything. Then Marvin decided that he wanted some fame and we got an article in *Time* magazine about Marvin and toy invention and what he did and how lucrative it was. And anybody who read that thought: 'Wow, I could become a toy inventor.' That really opened it up for the people that wanted to come and invent toys and present it to manufacturers."

But Glass's success was not just a product of his ego but also of his childlike view of the world. As the man himself told the *Chicago Tribune* in 1961: "You must see things as a child does—magically."

# 11

# SEX IN A BOX

## *What board games from Twister to Monogamy tell us about sexual attitudes*

Reyn Guyer's heart sank as Milton Bradley's Mel Taft broke the news. They were going to discontinue Twister. It was Christmas 1965 and the board game Guyer helped create had only just been released.

Twister was novel—a game where the players became the playing pieces on a large polka-dot vinyl mat that spread across the floor. Players would twirl the spinner and then contort themselves so that they could get a hand or a foot onto the colored dot it pointed to without falling over. It was a silly game but all the better for it. People would get tangled up and then collapse into a giggling mess of muddled limbs.

It's the retailers, explained Taft. Sears Roebuck had decided the game was too risky to carry.

Back then Sears Roebuck was the foremost retailer in the United States. The department store chain employed nearly one in every two hundred working Americans and a third of U.S. adults had a Sears Roebuck credit card. Sears Roebuck stores were the foundation of every successful shopping mall.

By refusing to put Twister in its stores and mail-order catalog, Sears Roebuck had all but guaranteed that the game would be a flop. So Milton Bradley cut its losses. It halted production of the game and took Twister's television ads off the air.

As he put the phone down, Guyer felt crushed. Just a few weeks earlier everything had felt so positive but now Twister was finished.

Born and raised in St. Paul, Minnesota, Guyer had wanted to be a writer. He studied English at Dartmouth College and after graduating in 1957 intended to become a newspaper or magazine journalist. It never happened.

Instead on returning to St. Paul he was, he says, "coerced" into joining his father's design agency, which specialized in creating end-of-aisle store displays for the products of Fortune 500 companies. "So I stepped up to the drawing board, having no idea what I was doing, and began to watch and learn," he remembers.

Six years on Guyer was still there. He had risen up the ranks to become a co-owner of the agency and its Midwest representative, working with many of its biggest clients. But he was also fed up with end-of-aisle displays.

"I was constantly looking for another way to broaden the activities of our company simply because the model that we'd developed I was not very fond of and didn't see the long-term growth potential in it," he says. "I kept looking for something else we could do that would be a little less dependent on having to come up with a brand-new idea for our clients every year."

One client Guyer looked after was the S. C. Johnson Company, a manufacturer of floor wax, bug spray, and other consumer chemicals. The company wanted Guyer to develop a back-to-school campaign for its new shoe polish and so he began thinking about creating what marketing people call a "self-liquidating premium." This promotional trick involved creating a "free" gift that customers might want but could only get if they sent in enough proofs of purchase along with a nominal fee for postage and packaging. If all went to plan, the extra sales generated by people trying to get the free gift would pay for the entire promotion while entrenching customers' loyalty to the brand.

The Pepsi Stuff campaign, which started in 1996 and continued into the 2000s, was typical of these promotions. During the promotion Pepsi drinkers who sent in their bottle caps would earn Pepsi

Points that could be spent on gifts ranging from Pepsi-branded hoodies to mountain bikes. But earning enough points to get any of these items required guzzling down plenty of cola. A Pepsi T-shirt, for example, cost eighty points, which equated to buying forty two-liter bottles of Pepsi-Cola.

Naturally the starting point for any self-liquidating promotion is to decide what gift to tempt people with. Guyer thought about making some kind of game, maybe something that involved children standing on a grid of squares. He toyed with this vague idea for a while and then had a flash of inspiration: what if the children had to move from one place on the grid to another?

Guyer rushed over to where the area of the office where agency's artists and designers worked and pulled out a large piece of the corrugated fiberboard that the company usually used to create its product displays. He cleared a space and laid the eight-by-six-foot board on the floor, grabbed some pens, and set to work. "I drew on quickly twenty-four squares in four-by-six arrangement," he says. Guyer then colored in two of the one-foot squares in each corner of the board. Two yellow, two green, two red, and two blue.

On finishing, he went around the office rounding up artists and assistants. The goal, Guyer told the bemused employees as they stared at his hastily constructed game board, is to move from your team's starting square to the colored square in the diagonally opposite corner. Each of us will take turns and move one square. Let's start.

The game was ridiculous. Players shuffled from space to space with their legs pressed together and soon found themselves bunched up uncomfortably close to one another and unable to move. Within ten minutes the game had dissolved in laughter.

"The game turned into chaos but we were laughing so hard it didn't make any difference whether we achieved the objective," says Guyer. "At that moment I knew I was onto something new and very different."

Encouraged by that early test Guyer began honing the game into something more structured. After a few weeks of experimenting he came up with King's Footsie, a game played on a white plastic sheet divided into twenty-five one-foot squares.

"It basically was four people being on the board at the same time, two teams of two, the red team and the blue team," he says. "Each of the players on the red team had red bands tied around their ankles and the blue team did the same. The first team to get all four of their feet in a row won the game. A four-way tic-tac-toe, if you will."

Convinced that King's Footsie was too complex for a shoe polish promotion, Guyer got in touch with another of the agency's clients, Scotch Tape, which was part of the 3M business conglomerate. 3M had recently started a division called 3M Bookshelf Games that published board games in boxes that resembled hardcover books. Guyer hoped 3M would buy King's Footsie, but while the company's game team liked his creation they felt it didn't really suit their upmarket image.

After being rejected by 3M, Guyer was stumped. He had no experience in the games business and had no idea how to go about selling his game to publishers. So he stuck the King's Footsie board on a shelf in the agency's conference room and refocused his attention on the day job.

Several weeks later the agency's purchasing director Phil Schafer held a meeting in the conference room with Chuck Foley, a salesman for a silk-screen printing company. After they finished talking business, Foley noticed the King's Footsie mat.

Schafer explained that it was a game Guyer had designed but wasn't doing anything with. Foley, who had worked in the toy business, asked to meet Guyer. After the two were introduced, Foley explained that until recently he had been working for Lakeside Toys in Minneapolis and so knew the games business well.

"Our conversation led toward, well, you do know some things about toy and game manufacturing that we don't and we need someone who is more familiar with that industry because we've got the idea but we don't know where to go or whom to see," says Guyer. "I was running the company at that time so I was very pleased to find someone I could team with to move my idea forward."

Guyer hired Foley and Foley's artist friend Neil Rabens and the three of them began developing games based on Guyer's "people as the playing pieces" concept. The three imagined creating a whole line of "stocking-feet games" to sell to publishers.

One of the team's first creations was a reworked version of King's Footsie. Foley replaced the grid of squares with rows of different-colored circles. The team then created a spinner that dictated what color circle players had to put their feet on. Finally, Rabens hit on the idea of getting players to use their hands in the game as well. They called the new game Pretzel.

By early 1965, with eight stocking-feet games ready, Foley arranged a meeting with Milton Bradley vice-president Taft in East Longmeadow, Massachusetts. Once they got into his office the team began demonstrating their games one by one. Taft was polite but clearly unmoved by what he saw.

Then they laid Pretzel's mat on the floor and began playing. As they tangled themselves up on the office floor, Taft perked up, and when the meeting was over he asked if he could keep Pretzel so he could pitch it to his bosses. A few weeks later Taft called Guyer to tell him that Milton Bradley wanted Pretzel.

Guyer and his team were delighted. The only disappointment was that Milton Bradley was going to rename it Twister because a plush toy was already using the name Pretzel. Guyer wasn't at all keen on the name Twister: "Having grown up in the Midwest of the United States and having been chased around by twisters throughout my life, I felt that the name Twister had a very negative inference to it."

Even though everything seemed to be going fine, back in East Longmeadow all was not well. Taft had fought hard to persuade his bosses to buy Twister and many of his colleagues thought a game like that had no place in the catalog of a family-friendly company like Milton Bradley.

The problem was sex.

The sexual revolution was only just beginning in early 1965. State laws criminalizing the pill and other forms of birth control wouldn't be overturned by the Supreme Court until later that year and, even then, only for married couples. The unmarried would have to wait another seven years before the Supreme Court confirmed their right to contraceptives.

And in a society where people kept what happened in the bedroom in the bedroom, and major movie studios adhered to the no-sex rules of the Hays Code, a game that involved people rubbing up against each other and getting into unusual positions on their living room floors seemed borderline indecent.

"We broke the rule that says that you really shouldn't be in proximity to another person in a social setting," says Guyer. "It really was not acceptable except when dancing with a person of the other sex."

So when Milton Bradley's sales team tried to get stores to stock Twister it was an uphill struggle. Retailer after retailer refused to take the game and when Sears Roebuck snubbed Twister for being too risqué, Milton Bradley gave up.

Yet, unknown to Guyer and Milton Bradley's management team, Taft left the game with one last shot at success. Before halting production, Milton Bradley had paid a public relations firm to promote the game and, since the money had already been spent, Taft figured there was nothing to lose from letting the media campaign run its course.

And in early 1966 the public relations team pulled an enormous rabbit out of the hat: an appearance on *The Tonight Show* with Johnny Carson, a television talk show watched by around ten million people every night. On May 3, 1966, Taft and public relations consultant Ruth Miller joined the show's audience in New York City. Neither had any idea what the show's unpredictable host would do with the game. For all they knew Carson might castigate Twister live on national television and finish the game for good.

It went better than they could ever have imagined. During the show

Carson's guest, the glamorous actress Eva Gabor, enticed the host to join her for a game of Twister. The sight of Carson in his suit and Gabor in her low-cut dress tying themselves in knots on the mat caused the live audience to burst out laughing and caught the rapt attention of millions of viewers.

The next morning people across the America went in search of Twister. In Manhattan people descended on Abercrombie & Fitch, one of the few stores that agreed to stock Twister, and began queuing down the block to buy it.

Milton Bradley quickly restarted production and relaunched the TV advertising campaign. "When Mel Taft called me and told me they had changed their minds it was a very happy moment in my life," says Guyer.

The prudishness about Twister might not have disappeared—one competitor even accused Milton Bradley of peddling "sex in a box"—but by the end of 1966, nearly five million copies of Twister had been sold.

If people thought Twister was improper it clearly wasn't doing the game any harm, and most people saw it for what it was supposed to be: a fun party game. As Foley later said of those who associated Twister with sex when asked by *Timeless Toys* author Tim Walsh: "Dirty mind, dirty game. Clean mind, clean game."

Still, it is hard to imagine Twister's up-close-and-personal action would have overcome the "sex in a box" accusations back in the frigid fifties. The timing of its release mattered. It arrived just in time to ride the wave of sexual liberation that would soon shake off centuries of prudishness and allow a book like Alex Comfort's *The Joy of Sex,* an illustrated guide to lovemaking that gave the thumbs-up to swinging, to spend more than three hundred weeks on the *New York Times* bestseller list during the 1970s.

And as people loosened up about sex, a rush of new amorous board games soon made a mockery of anyone who thought Twister was raunchy. These new games *wanted* to be sex in a box. There were games like Bumps and Grinds, "a hilarious drinking and stripping game" according to the box, and Office Party, the game where "all players really get down to the nitty gritty."

Boldest of all was Seduction, "a swinging game for swinging couples," which came in a box that depicted mustached men gazing lustily into the eyes of female party guests while sipping pale orange cocktails. Seduction based itself on Monopoly and sought to get players frisky by letting them buy properties with names like Stud Valley Dude Ranch and taking cards that issued commands like "kiss the player on your left" and "remove an article of clothing." According to *Today's Health* magazine Seduction proved so effective that it sparked an "ugly domestic" when one Californian man tried seducing his best friend's wife over the game board.

The salacious seventies, however, came to an abrupt halt in the 1980s, when the terrifying rise of HIV and AIDS poured a bucket of ice-cold water over the licentiousness of the previous decade. If the seventies was all about the joy of sex, the the eighties was the decade of Russian roulette sex, and the sexual board games of the time reflected that sudden shift by switching from wife-swapping to safe-sex education.

Leading the charge of the new safety-first sex games was Dr. Ruth's Game of Good Sex, a game endorsed by sex therapist Dr. Ruth Westheimer, who had become a household name thanks to her taboo-busting call-in radio show *Sexually Speaking*. Her 1985 game saw players trying to accumulate arousal points by answering questions about the cause of genital warts and how to wash semen stains out of fabrics. "It fills a function in an enjoyable way," Dr. Ruth told the Associated Press ahead of the game's launch, which made it sound about as titillating as doing the dishes with the radio on.

The education-before-eroticism theme continued in SEXploration! a 1987 game from the Cowell Student Health Center at Stanford University in which players could catch herpes or end up on a date without a condom. "We felt some students were making sexual decisions on an impetuous or emotional basis without thinking of the consequences beforehand," cocreator John Dorman informed the *Day* newspaper.

What both games reflected was how the specter of AIDS drastically altered sexual attitudes in the space of just a few years. The anything-goes zeitgeist of the seventies had given way to a world where disease and desire were intertwined.

Today the pendulum has swung back in promiscuity's favor, leading to games that focus on the pleasure of sex once more. But this time there are limitations. A 2015 study reported that while U.S. millennials, those born 1982 to 1999, are the most sexually permissive generation yet, they also have fewer sexual partners than the Generation Xers who preceded them.

So while sexual board games are more widespread than ever before, in line with the generational shift in attitudes toward sex, most stop well short of Seduction's partner exchanges and prefer to focus on encouraging couples to expand their sexual horizons.

This outlook is embodied by one of the most popular adult games on sale today: Monogamy, which seeks to add some zing to couples' love lives and has sold close to a million copies since its release in 2001. The game's British cocreator Jane Bowles got the idea for Monogamy after becoming a mom. "We had our two lads and when you've got two kids sex goes lower down the agenda and we would just sort of collapse in a heap in the evening and not do much," she recalls.

"I was thinking of ways we could spice things up and rather than read a book, which can be a bit embarrassing and a bit awkward because you've still got to suggest stuff to your partner, I thought wouldn't it be great if you could have a game that really helped spice things up and got you communicating again, because a game takes all the embarrassment out of it."

The trouble was the sex-themed games on sale were, she says, cringeworthy. Most were nothing more than smutty versions of Monopoly. Even the ones that didn't copy Monopoly were terrible, including Foreplay, a game that Bowles and her husband, Richie, distributed through their adult products distribution business Creative Conceptions.

Oddly the couple discovered Foreplay in a shipping container stranded on a farm in Devon. "A distributor had brought it into the country and never ended up paying for it, so it was stuck on a farm and somebody told me about it," she says. "Foreplay was kind of cringey and embarrassing. It certainly didn't make your tummy tingle or anything. It was rather like a medical textbook in its approach. But we parceled it up and it sold well, so it showed there was a market for a game for two that does work at relationships."

Faced with the demands of parenthood Bowles wanted a better game, one that went beyond the usual "shake the dice and get it on" fare. "We wanted it to be a total relationship game that helped you remember why you are together in the first place," she says.

The result was Monogamy. In the game players travel the board's circular track, landing on spaces and taking cards that suggest activities designed to heighten the mood, and as the game progresses the actions get steamier and steamier.

Monogamy starts at its "intimate" level with kisses on the cheek, nibbles of ear lobes, recollections of players' first sexual encounters, and shoulder rubs. Mid-game it develops into players telling each other about what they used to find a turn-off but now fantasize about, peeling off each other's clothes, and licking thighs. Then, in the final "steamy" stage, the game gets players fondling blindfolded partners, eating food off each other's naked bodies, pulling underwear off with their teeth, and more.

Players also collect Fantasy Cards during the game. These cards offer suggestions like making love in the shower or kinky role-playing scenarios and are designed to bring the game to a very literal climax, because the first player to circle the board six times gets to choose which of their fantasy cards to make reality.

Perhaps surprisingly for someone who made a sexual board game and founded a company that distributes vibrators and condoms, Bowles gets a little coy when asked where the ideas for the actions and fantasies in Monogamy came from.

"I hope our sons never read this," she says. "Much of the game came from our own thoughts and imagination. It was quite scary putting our fantasies in, especially when we were giving the game to friends at the start. You're thinking: 'Oh God, do they think this is all from what goes on in our bedroom?'"

In addition to Bowles and her husband's own imagination, the game also included suggestions from friends and ideas from research into what keeps relationships alive. And while Bowles might have felt slightly embarrassed when asking friends for feedback on the game, it was their reactions that convinced her Monogamy could appeal to the wider world. Not least because the game had life changing results for one of their play testers. "Richie gave it to one of the girls in his team and she ended up getting pregnant," she laughs. "So we kind of got the feeling that it definitely worked."

What makes Monogamy work, says Bowles, is that it's a board game. "The actual act of sitting down with a board game does a couple of things," she says. "I think it evokes good memories because board games are, when you look back over your childhood, usually fun times of getting together and talking.

"Also because it is a board game there are no screens involved so it does make you stop and concentrate and focus on what you're doing. You're not being distracted here, there, and everywhere by things bobbing up on a screen. You just sit down and focus on each other."

Monogamy would have been unthinkable back in 1965 when Twister launched, yet even now—in an age of Victoria's Secret, *Fifty Shades of Grey,* and sex toys on sale in CVS—retailers still balk at the idea of sexually themed board games. When Creative Conceptions first tried selling Monogamy to British chain stores they were turned away, and more than fifteen years later nothing's changed. Even the huge success of *Fifty Shades of Grey* hasn't changed retailer attitudes.

"You still get that objection from mainstream stores," she says. "I remember being in Tesco [the UK's leading supermarket chain] when *Fifty Shades of Grey* arrived and it's on aisle ends and it's not sealed and anybody of any age can open up the book and see some quite hardcore language. But because Monogamy is an adult game, we've always had resistance even though absolutely nothing on the box or back of the box would cause offense."

Because of this Monogamy has been a below-the-radar success, steadily building sales through word of mouth and online stores to a point where it had even been spoofed in an episode of *Family Guy* as "Hot Monogamy, the board game for failing marriages."

As it happens Monogamy has been used by the British relationship counseling charity Relate to help couples who are struggling with their sex lives. "We don't like to pigeonhole it as something that is for people who are struggling in relationships because it's not; it's very much something that anybody in any relationship could play and have a really good time," says Bowles. "But there are some Relate people who give the game to their clients."

Since the launch of Twister board games have echoed Western society's sexual journey. From the uptight attitudes to Twister in the sixties, we have moved through the permissiveness of the seventies and the biology lessons of the eighties to arrive at games that embrace sexually adventurous monogamy.

Yet while our view of sex and the flirtatious board games embodying those attitudes have changed greatly in the past half century, Monogamy's struggles to get on store shelves despite strong online sales suggest that when it comes to retail at least, maybe attitudes have changed less than we think.

12

# MIND GAMES: EXPLORING BRAINS WITH BOARD GAMES

## *What board games reveal about our minds*

From the moment she was born, Judit Polgár was to be an experiment. Even as she took her first breath in a Budapest hospital on July 23, 1976, her life had been mapped out.

Her parents had long ago decided that she would be a chess grandmaster. It may have sounded like the talk of parents who hoped that their newborn child would live out their own unfulfilled fantasies, but her educational psychologist father László Polgár was out to prove a point. He believed that geniuses are made, not born, and was going to use his children to prove it to the world.

After his wife, Klara, gave birth to their first daughter, Zsuzsanna, in 1969 they began debating what subject they should train her to master. Should she become a genius mathematician or a master of foreign languages? They were still undecided when Zsuzsanna began toddling around their modest apartment in downtown Budapest. Then four-year-old Zsuzsanna opened a cabinet drawer, found a chessboard inside, and asked to be taught how to play. In that moment the question was answered. Zsuzsanna, who later renamed herself Susan, would be a chess prodigy. Her parents devised a home-schooling program that involved tutoring from expert chess players and would see their daughter drilled in the art of chess.

In 1974 their second daughter Zsófia was born and she too became part of her father's grand chess experiment. And when their last child Judit was born it was inevitable that she too would join her sisters at the chessboard.

Judit's training began at the age of five. "I remember when I started to learn the moves from my mother, learning the moves each piece at a time," she says of her first memories of the game. "It's only one piece at a time, jumping all over and moving that piece around the board to get stable knowledge about that piece."

Chess dominated the three sisters' childhoods. They would spend at least five to six hours every day honing their skills and studying the games of history's greatest players. "In the family everybody was playing chess," says Judit. "By the time I started to play chess, it was a regular routine. It was very natural that I would be following the education of my sisters and that I will also not be present in school."

Her parents' decision to home-school their children did not sit well with Hungary's communist government. "My parents had a lot of difficulties," says Judit. "In those times when people had individual ideas different from the regular, normal way, that was not welcome at all. There was a point where the government was considering that they would take my parents to jail and take the kids to the orphan house."

The sisters' intensive schooling in chess quickly produced results. Even before Judit was born, a four-and-a-half-year-old Susan had challenged adults in Budapest's smoky chess clubs. Many of those who humored the little girl soon found themselves humiliated and shaking hands with a victorious child.

All three sisters would become great chess players. Susan became the world's top female chess player in 1984 at the age of fifteen and in January 1991 she became the first woman to achieve grandmaster status on the same basis used to assess men for that title. A fourteen-year-old Zsófia, meanwhile, stunned the chess world at a 1989 tournament in Rome by defeating a string of Soviet grandmasters in what became known as "The Sack of Rome."

But it was Judit who shined brightest. By her sixth birthday she could already beat her father and soon after she began competing in local tournaments. A year later she was taking on adult masters while blindfolded.

At nine she went to New York City to play in her first international tournament and won in the unrated category. "It was a special feeling, you can imagine, for a nine-year-old girl to beat an adult, or even to be in competition," she told CNN.

Not that taking on adults fazed her. "For me it was very natural to play against adults and be successful but, of course, for the opponents it was not fun to play against a little girl and possibly have a very tough game or even lose the game," she says. "Some people behaved very badly, others took it as interesting and something positive."

In 1987 an eleven-year-old Judit defeated her first international grandmaster and the following year she became the youngest player to become an international master.

At age fourteen Judit abandoned the women's tournaments and moved into the male competitions. The chess establishment, which regarded women as inferior players, was far from keen on the Polgár sisters' aspirations to cross the gender divide. In 1986 when Susan, then seventeen, qualified for the Men's World Championship, the world chess federation FIDE refused to let her play.

But the Polgárs' abilities and determination smashed through the chess world's gender barriers like a wrecking ball. Confronted with these exceptional prodigies, the chess authorities had little choice but to change their policies. "Before it was called the Women's Olympiad and the Men's Olympiad," says Judit. "Now they have changed it because of us. Because of the three sisters it's now the Women's Olympiad and the Open Section."

In the years that followed Judit became a star within the confines of competitive chess. She took on many of the greatest and most famous players who ever sat before a chessboard and won. Magnus Carlsen, Anatoly Karpov, Garry Kasparov, Boris Spassky: Judit scored victories against them all.

By the time Judit retired from competitive chess in August 2014, she had etched her name into the annals of chess history as the greatest female player of all time and the sisters had proved their father's theory that geniuses are made, not born, in spectacular fashion.

The Polgárs were not the first to use board games to throw light on the workings of the human mind. For that we need to go back to the 1890s, a time when chess masters would stun crowds by playing as many as ten simultaneous games while blindfolded.

These grand displays of memory and analysis impressed Alfred Binet, a young psychologist living in Paris at the time. Interested in the nature of human intellect, Binet wanted to understand how these chess players mastered the game. So he arranged to interview the French master Alphonse Goetz, who had recently played eight games at once while blindfolded.

The popular consensus was that these blindfolded players could achieve such feats because they had exceptional minds that allowed them to construct a detailed and complete picture of the boards and the location of every piece in their mind's eye. But what Goetz told Binet took the young psychologist by surprise. There was no mental photograph. Confused by this unexpected revelation Binet invited a group of blindfold chess masters to his laboratory for deeper questioning.

What he learned from them confirmed Goetz's revelation.

When Binet asked the players to sketch how they imagined the game in their mind, they drew a ghostly, fuzzy board that was missing most of its squares. The sketches were utterly devoid of playing pieces too. Instead there were vertical, horizontal, and diagonal lines that echoed the potential movement of their own and their opponent's pieces. Visualizing more would just be a distraction, explained the players, everything of importance is in the sketch. There was no need to know the color of the squares or the shape of any piece. All that mattered was is there a piece on a square and what directions might it move.

As Goetz told Binet: "I do not see the shapes of the chessmen at all. . . . I am aware only of the significance of a piece and its course. . . . To the inner eye, a bishop is not a uniquely shaped piece, but rather an oblique force."

In contrast, when Binet asked amateur players what they envisaged when playing blindfold, they said they tried to imagine the exact board with no detail overlooked. The chess masters with their deep understanding of the game knew what information to focus on while their amateur counterparts had to construct a complete picture in their mind to work out their moves.

Binet's experiment would only be the first of many that sought to unpack the workings of the human mind by comparing grandmasters with novices. Chess proved so useful in studies of the mind that researchers began calling the game the drosophila of cognitive psychology, a reference to the fruit flies often used in genetics experiments.

In 1973 experiments with chess led William Chase and Herbert Simon, two psychologists from Pittsburgh's Carnegie Mellon University, to develop a theory of memory called "chunking," an idea that came to be regarded as one of the most important in psychology. For their research Chase and Simon tested the memories of three groups of people: chess masters, experienced players, and novices. They gave their subjects brief glimpses of different board positions before asking them to re-create what they saw. They also tested their subjects' recollection of each move in a full game of chess. The chess masters comfortably outperformed the other groups, and this, Chase and Simon proposed, was because of chunking, the process by which we break information down into easier-to-remember pieces.

The classic example of chunking is how we remember phone numbers, or at least did until we got cell phones. Memorizing a ten-digit sequence of numbers such as 2, 1, 2, 5, 5, 5, 2, 3, 6, 8 is hard. So we break it into chunks, turning it into three groups of numbers like 212-555-2368, which makes it much easier to recall. It's not unlike eating. Instead of trying to shove a whole hot dog down our throats, we bite it into manageable chunks that we can swallow.

The chess masters, the theory argues, have chunked information from thousands of different chess situations, openings, endgames, and formations. They learn to associate particular board configurations with possible tactics and strategies. It's almost like a language where the chess pieces form the alphabet and their arrangement on the board are words infused with their own specific meaning. It's not for nothing that Judit Polgár calls chess her "second native language."

By chunking their chess knowledge, expert players could recall the situations they were shown better than nonexperts because they had already memorized similar circumstances and understood the meaning of what they were shown more deeply than lesser chess players.

To prove their point Chase and Simon then asked the three groups of chess players to recall meaningless and unrealistic arrangements of pieces on chessboards. This time the chess masters, unable to relate what they saw to what they had learned, performed no better than anybody else.

MRI scanners have since added weight to the theory. One study compared the brain activity of expert and inexperienced players when playing chess. The inexperienced players showed more activity in the hippocampus and medial temporal lobe, areas of the brain linked to short-term memory. The experts, however, had more active frontal lobes suggesting that they were drawing on existing knowledge of the game.

While chess was helping dissect the processes of memory other researchers were trying to find out whether chess and intelligence were connected. There was certainly no shortage of claims about chess's educational and intellectual value. Indeed chess is often discussed in a way that makes it sound like the board game equivalent of eating your greens, but was mastery of it really a sign of superior intellect?

In 1925 a team of Russian researchers tried to answer that question by putting top-rated chess players and novices through a series of psychological tests. They found that the only area in which the chess experts excelled was chess, a finding later supported by many subsequent studies. Chess masters, their results suggested, have exceptional knowledge about the game but not necessarily exceptional brains.

As for the related question of whether chess benefits children's education, the evidence is largely in the game's favor, although the significance of the reported gains varies greatly and some studies have found no noticeable difference in the progress made by kids who play and those who do not.

Among the apparent educational advantages of playing chess are improved concentration and an enhanced ability to plan ahead. Some studies report that chess-playing kids do better on reading, math, IQ, critical thinking, and visualization tests.

It's not only children who may benefit from playing the game. There is also evidence that the mental stimulation chess offers can help the elderly reduce their risk of developing Alzheimer's disease. Whether these qualities are unique to chess or true of any skill-based game, however, is unknown.

One commonality between all these studies is that they associate chess with the analytic, intellectual, and rational mind. Chess can tell us about how we feed and use our memories, and how we might visualize things inside our heads, but what about the irrational mind? Can board games reveal anything about our emotional nature?

On the emotional front the logical realm of chess offers few—if any—revelations, but an altogether different game does: the Ungame, the creation of an emotionally repressed Californian housewife that lured millions of people into opening up.

The roots of the Ungame began in the buttoned-up childhood of its inventor, Rhea Zakich. "I grew up in a family that never ever acknowledged any emotions," she recalls. "No hugging, no kissing, no touching. We weren't allowed to cry. I would hear terms like don't be a sissy, don't be a baby, and that's nothing to cry about, and so I really grew up thinking that I was to outgrow my feelings."

As a child of the 1930s and 1940s, Zakich's experience was far from unusual. "Feelings didn't have names because nobody talked about them in those days," she says. "We had just had the Great Depression. In those days life was hard and feelings really didn't have anything to do with it. It's not like you wanted to have a job that gave you great pleasure, it's like if you had a job you just were glad. People didn't search for pleasure so much—it was just survival."

So whenever she felt emotions Zakich pushed them away, and her sense that emotions were a shameful weakness stayed with her into adulthood. Repressed emotion after repressed emotion piled up inside her.

Then in 1970 several polyps were found on her vocal cords and the thirty-five-year-old mother of two had to undergo surgery to remove them. After the operation doctors told Zakich not to talk or make a sound for months because even the slightest sound could result in bleeding that would cause her to lose her voice for good.

The imposed silence left her isolated. "When I was a mute nobody paid attention to me," she says. "People didn't know what to say to a person who can't talk so they would just put their head down and walk past."

Her children and her husband stopped talking to her. They would come home, eat dinner, watch television, and talk to each other as if she were invisible.

Zakich began thinking about what life would be like if she never got to speak again. "I was haunted by the fact that most of what I had said to my children were directions," she recalls. "Don't do this and do that and do your homework, wash your hands, feed the dog, take out the trash. I became aware that everything that came out of my mouth was a barking order for my children. So in my muteness I was certainly aware of the regret of forgetting to tell them how much I love them and cared for them and why I was being strict."

She felt the same about her marriage. "We had become so businesslike," she says. "My husband would come back from work and I'd say: 'Darren needs a new pair of shoes and the screen door is still squeaking and I need to go to the store tonight.' Nothing kind and loving because it was all business. Why didn't I ever think to say I like him or that I'm sorry or that I cared? I forgot all those things in the business of it all."

The upswell of emotions made her feel like a shaken can of soda pop that was about to explode. "I thought if anything else happens I would literally explode like a bomb," she says. "So I thought, 'I'm going to let people know that I have all this going on inside me.' In a way I wanted to blame everybody else."

So she began writing cards with the questions that she wished her family would ask her about how she felt, hoping it would give her an excuse to let the emotions pour forth.

As she wrote them, Zakich wondered how on earth she was going to get her family to take notice of her questions. "Then I thought: 'Ah! We used to play games at the kitchen table,'" she says. "So I spent a day, while they were at school and my husband was working, drawing a game board, just making it up—a little road and little sections. Then I got the pieces from our Monopoly game and I set them on my little game board and put the deck of cards in the middle of the board and I just let it sit there."

Her game was an unusual one. There was no goal, no winning, and no losing. The track the playing pieces moved around was an endless loop. As they went around the board players would pick up cards from the deck and respond to the questions. The only rule was that players must listen to the person answering a question from the card deck without making comments or asking questions.

Worried that too many questions about feelings would put her family off the game, Zakich added fun and throwaway queries like "What's your favorite color?" and "What do you want for your birthday?" She then carefully arranged the card deck so that she would get the questions she wanted to answer when—if—they played it.

On returning from school one of her sons spotted the handmade game on the kitchen table and asked her if they could play it after dinner. Zakich nodded. It was the first time he had spoken to her in days.

After dinner Zakich and her family gathered at the table for the game. Then, just as they were about to begin, one of her sons grabbed the deck and, to Zakich's horror, shuffled the cards. Now there was no telling who would be asked what. "I thought, 'Uh-oh, I'm not going to get the cards I wanted, they are going to get them,' but that was the amazing thing," she says.

The next twenty minutes changed the family forever. Her husband drew a card that got him talking about how lonely he felt and how he missed the sound of his wife's voice. "I didn't know he was ever lonely, he told the story, and the boys listened," she says. "Then my younger son drew a card that said: 'How do you feel when people laugh at you?' I wanted to answer that because people did tease and make jokes like, 'Did you ever think you'll see Rhea with her mouth shut?' which did not strike me as funny at the time.

"So my son gets this question and he says: 'I hate it, I hate it, these kids tease me all the time, they make fun of me and call me fat and sometimes I think I will make myself dead so they will feel sorry for me.' I was totally shocked, we all were. He had these thoughts and we had no idea."

Soon after, her other son, a straight-A student, was telling them how he lived in constant fear that the rest of the family wouldn't like him as much if he got a low grade. "In less than twenty minutes we were sharing things with each other that we had never shared before," says Zakich.

A few days later her husband invited their neighbors round to play the game. They played for just fifteen minutes but at the end of the evening, the neighbors asked if they could take the game home with them. Our kids hardly talk to us anymore, they explained. Thrilled that they liked her game, Zakich nodded and then realized that she had just given away the object that had enabled her to end all those years of emotional silence.

So the very next morning she set to work making another copy. Over the next six months Zakich found herself constructing dozens more, since everyone who played her game seemed to want a copy. Those who got a copy introduced the game to others, who then in turn borrowed it and introduced it to even more people. Soon Zakich was getting letters from people she had never met who had played her game and wanted her to make a copy for them. Then came letters from church pastors and schoolteachers who wanted multiple copies for their youth groups and classes.

Her days became filled with the task of typing questions cards and drawing game boards that she would color with crayons. Eventually people were writing to her not only to ask for copies of the game but also to explain how it changed them and their families for the better.

Zakich began thinking she should sell the Ungame and wrote to game manufacturers and educational supply companies.

Every single company she wrote to wouldn't publish it. It won't work, they said. People don't talk about their feelings, there has never been a game about emotions, and, besides, since no one can win it's not even a game. "No one had ever heard of a noncompetitive game or one that had to do with serious topics," she says. "I didn't want it to be competitive, I didn't want anyone to lose after sharing their feelings."

Then in 1972 her luck changed. A boy from her neighborhood had found a half-finished copy of the game that Zakich had chucked in the garbage after making a mistake. Shortly after the boy's parents approached her and said they were willing to remortgage their house to help put it into production. "It had to be some kind of divine miracle because the people in the business did not think it would work—games were about pretend and this was about being real," says Zakich.

Together they founded the Ungame Company and set about trying to get stores to stock the game, but since they lacked the money to pay for advertising, retailers were reluctant to put it on their shelves. How will anyone know the game exists? they asked. So Zakich, who by now had recovered her voice, began giving talks about the Ungame at schools around Southern California and alerting local toy stores that people would be looking to buy the game afterward. Slowly but surely retailers began taking copies, selling out and then ordering more.

From there the Ungame's momentum grew and grew. A few dozen sales became a few thousand, which then became tens of thousands, then hundreds of thousands. By 1985 sales had passed the million mark and the game was still selling.

Zakich found herself with boxes of letters from people all over the country. Countless tales of saved marriages, uncommunicative children brought out of their shells, mental health patients who opened up to their therapists, and accounts of how playing the Ungame helped families cope with the death of a loved one.

The Ungame chimed with the zeitgeist of the 1970s, a decade where the idea of getting in touch with your emotions rather than batting them away took root. The newspaper advice columns, therapists, and self-help books that once told people to control their emotions were now urging them to open up.

As society's attitude to emotions changed and sales of the Ungame rose, similar games with names like Group Therapy and the Talking, Feeling, and Doing Game also began to get traction.

Some psychologists and psychiatrists fretted about the trend, warning that these games would leave people exposed and vulnerable. "I'm familiar with the game, Group Therapy, and it appalls me," psychologist Dr. Edmund Shimberg of Lankenau Hospital told the *Philadelphia Inquirer* in 1972. "It plays with people's emotions, opening them up wide with no one to close them up afterward. It leaves their emotional guts open for friends and neighbors to see."

Other therapists, however, integrated these games into their work and praised them as useful tools for getting people to talk about their innermost feelings.

For Zakich the power of the Ungame to encourage people to express themselves lies more in the board than the questions on the cards. "When people sit around a game board they are sitting closer than they would be otherwise," she says. "When you sit around the game board you are huddled and that's part of sharing secrets. You get into a circle close together. You don't have to project your voice."

Equally important is the rule requiring players to listen without comment. "People get used to not talking unless it's their turn and so they are listening at a different level; they are not rehearsing in their heads what they are going to say," she explains. "So if someone shares their sadness because their brother died you can see the tears in the eyes of the other people. If it was just an ordinary conversation, you hear about the death of somebody's brother and you're thinking about the death of your aunt and your head is used for other thoughts."

Zakich may have stumbled upon board games' potential as a life-changing device but on Court Street, Brooklyn, a more deliberate attempt to use games as a way to improve lives is under way.

From the outside the Brooklyn Strategist looks like just another of the fashionable board game cafés that have opened in cities the world over in recent years. Inside there are bare brick walls, exposed water pipes, tables illuminated by funky metal lamps, and a coffee bar offering homemade sodas, lime coconut cookies, and Australian-style savory DUB (Down Under Bakery) Pies.

But the Brooklyn Strategist is founded on more than a love of games and good coffee. Its real foundation is neuroscience.

Dr. Jon Freeman is the owner of the Brooklyn Strategist. Prior to opening the café, he had built a successful career in clinical psychology that included directing a Manhattan neuroscience research lab and doing consultancy work for pharmaceutical companies, only to find himself feeling burnt out by his work.

So he began looking for a way out and found it at home under his TV set. "I was watching my seven-year-old daughter who would finish school for the day, and if she didn't have friends to engage with she would drift into this world of digital isolation," he says. "Whatever the digital toy de jour was—the Wii, the DS—she wanted it and would just so easily drift into that."

Freeman began thinking about how he could encourage his daughter to spend less time playing Nintendo and more time engaging with the world beyond the screen. It dawned on him that she really got engaged when they played board games together. "I thought, hmm, I can't sit and play board games with her every day . . . or can I?" he says. "And then I said if I am going to sit and play board games with her every day, could I generalize this into a larger program?"

He envisaged an after-school program based around board games that would help children develop in a fun way. He checked to see if anything like it already existed. It didn't. So he used his expertise in neuroscience and psychology to develop a program.

The twelve-week programs Freeman created aim to activate and develop particular parts of the brain through games that emphasize strategy over chance. The approach is based on the concept of neuronal plasticity, the idea that the neuron connections in our brains are constantly rewiring themselves in response to stimuli rather than, as once thought, reaching a static state on entering adulthood. "My programs are based on the theory that neurons that fire together, wire together; and the more active an area of the brain is, the more neuron-developmental growth you have in that particular area," says Freeman. "There's fairly compelling evidence in the neuroscience literature to support that idea."

Brain Benders is typical of the after-school programs Freeman has developed. Aimed at kindergartners and first graders, Brain Benders starts with games focused on pattern recognition and sequential learning before building up to games that demand linear thinking. The games used in the Brain Benders program range from the mancala game oware to recent releases like 2011's Mine Shift, a two-player twist on Chinese checkers.

Mine Shift uses a board created from ten tiles plucked at random from the box. Each tile contains four squares and differently positioned walls. As in Chinese checkers, players have to move their counters across the board to their opponent's starting area, but instead of just moving the pieces players also rotate and shift the placement of tiles to clear barriers to their own progress and block the way for their rival. "It's not a simple hop, skip, and jump like in Chinese checkers, it involves sequential rotation of tiles so they make connections," says Freeman.

After getting good at Mine Shift, the children go on to play games that help to introduce them to the concept of linear math. Kindergartners don't understand numbers in the same way as adults, explains

Freeman. While adults understand numbers in a linear way, young children understand them in a logarithmic way.

Let's say I draw an unnumbered line representing a scale from zero to one hundred. If I ask an adult to place counters with the missing numbers along that line, they will spread the numbers out evenly because they understand that numbers increase in a linear manner. But kindergarten-age children will place those numbers in a logarithmic way, where the lower numbers are spaced far apart and the higher numbers are clumped together at the end.

The shift from a logarithmic to a linear understanding of numbers happens around the ages of five and six, and the speed at which children make that conceptual leap seems to affect how well they do in math. "Brain Benders is about converting logarithmic to linear thinking in mathematics and, wouldn't you know it, once we started doing that with kids who were in kindergarten, I started hearing from parents that their kids' math scores started getting better," says Freeman, who hopes to back up these anecdotal signs of achievement with a proper study at some point.

The Brooklyn Strategist's programs are also having a positive effect on children who have special needs. Jennifer Gebhardt, whose nine-year-old boy Jason is a Brooklyn Strategist regular, recalls the effect its work had on one of her son's friends. "This kid was a very smart kid but just had a lot of special needs stuff going on," she says. "But when he was at the Brooklyn Strategist with Jason he could get into the game and calm the other parts of his behavior that were a little out of control and have a connection and friendship with Jason."

The Brooklyn Strategist has also helped to spot undiagnosed dyslexia and color blindness after staff noticed that those children struggled to move from nonverbal to verbal games or repeatedly identified the wrong playing pieces as theirs.

Another area of work the Brooklyn Strategist has become involved in is helping develop children's social skills, something that Freeman didn't anticipate having to do when he started the business.

"It's fairly clear that a lot of kids are coming in today with a social handicap based on digital communication," he says. "They really struggle with how to interact with people. They don't quite know how to talk to people, they don't know how to encourage other people, they don't know how to accept gracefully comments about their own behavior, and they don't know how to behave when something isn't going their way."

Many of the young people with poor social skills fall into what educators call the "2e" or "twice exceptional" category. "These are kids who are off the charts in some academic areas and not even on the charts in others," he says. "More often than not they are not even on the charts when it comes to things like socialization."

To address these children's underdeveloped social skills Freeman integrated role-playing games into his after-school programs.

"When I originally wrote my program, Dungeons & Dragons and role-playing games had no part of it whatsoever," he says. "Now we're running role-playing games every day and kids are building their own role-playing games. For the 2e kids in particular we use a game called Mouse Guard, which is a role-playing game that awards character points based on appropriate expression of affect. So you take on the role of a mouse and if your mouse is appropriately happy for another mouse or appropriately sad for another mouse, you get points and you build up your character."

Where the Brooklyn Strategist's child development work will lead, Freeman doesn't know. He hopes that he might get to open more stores so he can bring the after-school club to more neighborhoods, but that's as far as the vision goes for now.

"I have no grand plans," he says. "I feel like I'm getting younger doing this, I'm really enjoying it. It's the community that has made us as successful as we are. It's really quite synergistic. I was kind of surprised—starting this out I had no idea if the community would respond."

What's more, he adds, the Brooklyn Strategist's mix of after-school programs, ad hoc gaming, tournaments, chess nights, Scrabble competitions, special events, and themed nights has brought the whole community together.

"Ultimately what I think we represent is a psychologically and physically safe social space for people to come, whether it's kids or adults," he says. "If you look at our events you will see a real blend of kids and adults playing together, and in this society people are so freaked out about, you know, the idea of 'What's that adult doing talking to that kid? This can't be any good.' Well, in days of old when people socialized together, there was nothing wrong with that. Kids might have learned from their elders and maybe the elders would learn from the kids. I think we're sort of a throwback in that respect."

Its approach might be distinct, but the Brooklyn Strategist is also continuing the tradition of using board games to explore and enhance the mind. Like Alfred Binet in 1890s Paris, it is turning games into tools that can peer into people's minds. Like the Polgárs and school chess clubs before it, the Brooklyn Strategist uses games to enhance intellect. And, like the Ungame, it is bringing people closer together through play and teaching them to get along.

But board games' use in exploring the concept of intelligence doesn't stop with people, for, as we are about to find out, they are also paving the way for machines that think.

13

# RISE OF THE MACHINES: GAMES THAT TRAIN SYNTHETIC BRAINS

*How board games have powered the development of artificial intelligence*

Garry Kasparov entered the room almost swaggering with confidence. It was May 3, 1997, and the world chess champion was out to prove, yet again, that machines are no match for flesh-and-blood grandmasters.

Kasparov had been besting the world's finest chess programs for years. In 1985 he took on thirty-two chess programs at once and beat them all. In 1989 he slapped down IBM's Deep Thought chess computer in New York and in 1996 he had, despite losing the opening game of the match, done the same to Deep Thought's successor Deep Blue.

This time he was facing an improved Deep Blue and hoping to shame IBM's multimillion-dollar supercomputer for a second time from the comfort of a small New York City television studio as the cameras rolled. Kasparov comfortably won the first of the six games. "It's just a machine. Machines are stupid," he boasted afterward.

But beneath the bravado, something was bugging the Russian grandmaster. Late in that first game Deep Blue made an unusual move, nothing capable of stopping Kasparov's march to victory, but unexpected enough to rattle him. He pondered what it meant. Did this strange move hint at a long-term strategy the computer had in mind?

Was it a sign that there was more intelligence lurking in that fridge-size heap of microchips than he had credited it with?

Years later one of Deep Blue's creators revealed that the wayward move was due to a software bug that caused a move to be picked at random, but Kasparov didn't know this and the computer's strange decision ate away at his confidence.

In the next game things didn't go according to plan for the grandmaster. Kasparov found himself on the back foot and eventually he conceded defeat, having overlooked an opportunity to force a draw. The next three games also proved tricky for humanity's chess champion, each ending in a deadlock.

Now everything rested on the final game that was to take place on May 11. The looming showdown between man and machine captured the world's attention in a way that no chess match had since the Cold War clash between Bobby Fischer and Boris Spassky in 1972. "The future of humanity is on the line," declared one excitable *CBS News* anchor, before breezily moving onto the weather.

With the world watching, the May 11 match was tense. Kasparov grimaced and clasped his hands to his head as the emotionless Deep Blue punched holes in his defenses before forcing him to admit defeat after just nineteen moves.

Stunned, Kasparov rose from his green chesterfield office chair and quickly walked away from the scene of his defeat, arms outstretched and palms held upward in shock. The machine had won and humanity's days as the masters of chess were over.

The aftermath of Deep Blue's victory prompted plenty of soul searching. The media debated whether it marked a turning point in the relationship between us and our machines, the dawn of an age where intelligent computers would overshadow their creators. Others played the victory down, noting that Deep Blue couldn't do anything but play chess and so talk about computers replacing people was rather premature.

An embarrassed Kasparov accused IBM of cheating, claiming that Deep Blue must have had covert human help, and asked for a rematch.

IBM angrily rejected Kasparov's talk of subterfuge and refused the rematch, preferring to keep Deep Blue in the history books forever more.

Regardless of the debate about the meaning and nature of Deep Blue's victory, Kasparov's defeat was a momentous occasion for artificial intelligence research—the climax of a fifty-year quest to create a chess computer that could fell world champions.

The dream of chess-playing machines first took root in 1770 when Wolfgang von Kempelen unveiled his Automaton Chess Player at Vienna's Schnönbrunn Palace and left the court of the Habsburg Empire awestruck.

Von Kempelen's machine consisted of a large chest with a chessboard on top and the life-size torso of a mechanical man dressed in Turkish robes and a turban who moved the pieces around the board. The Turk, as it was nicknamed, played chess impressively well, taking on and usually defeating those who dared to challenge it.

After impressing Viennese high society, the Turk toured Europe, chalking up victories against Benjamin Franklin and Napoléon Bonaparte along the way. But it was all a con. The Turk's abilities were in fact the work of a skilled chess player who was operating the machine while hidden inside its large chest.

Despite the deception, the idea of chess-playing machines stuck, and in 1910 the Spanish inventor Leonardo Torres y Quevedo took up the challenge. The result was El Ajedrecista, an electric automaton that could play simple endgames of chess where a king and rook sought to trap a lone king. Impressive as El Ajedrecista was for its time, it would be the computer scientists of the 1940s who really got the quest to develop chess-playing machines rolling.

After the Second World War the fledgling field of computer science had two leading lights: the British mathematician Alan Turing and the U.S. electronics engineer Claude Shannon. The pair spent the war as code breakers, with Turing, in particular, playing a central role in cracking the German navy's secret Enigma codes. After the defeat of the Axis powers, the pair turned their attention to laying the foundations for modern computing.

Both saw the creation of an artificial intelligence as the ultimate goal of computer science and they agreed that getting a computer to defeat human chess players would be an important step toward achieving that aim.

Chess's complexity made it an appealing test bed for artificial intelligence research. While a child could learn the rules, the variety of possible situations in the game is vast. So vast that even if a computer played a million games a second, it would take far longer than the time thought to have elapsed since the Big Bang for it to process every permutation of the game.

This meant that a successful chess program would need to react to and anticipate the moves of a human opponent in some kind of intelligent way. As Shannon put it in a 1950 paper: "Although perhaps of no practical importance, the question [of computer chess] is of theoretical interest, and it is hoped that a satisfactory solution of this problem will act as a wedge in attacking other problems of a similar nature and of greater significance."

Chess became a holy grail for artificial intelligence research. The first proposed solution was to get computers to explore a "game tree," an ever-branching flowchart of possible moves. Starting from the current position of pieces on the board, the program would examine every possible next move before examining every move that could follow those moves, and so on until the end of the game. Along the way the computer would assess how good each potential move was and assign each one a score. Finally, having assigned scores to every move in the game tree, the computer would select the least bad move available. Computer scientists called this process "minimax."

While great in theory, minimax was slow. Running through all those moves required a lot of processing power, and processing power was in short supply in the computers of the 1950s. The only way to accelerate the process was to limit how many moves ahead the program would check, but the greater the cap on the computer's ability to look ahead, the worse it got at playing chess.

Clearly a faster approach was needed and in 1956 John McCarthy, the man who coined the term "artificial intelligence," came up with an enhancement to minimax called "alpha-beta pruning."

Alpha-beta pruning cut down the time computers wasted in assessing the outcome of inferior moves. Now instead of calculating every eventuality before making a move, chess programs would only look at options that promised to be better than the ones already assessed. With less processing time being wasted exploring fruitless moves, computers could now spend more time looking more moves ahead, boosting their chess playing prowess.

The breakthrough was so significant that some researchers got overexcited and declared that computers would be defeating grandmasters by 1970. Their predictions were decades out.

There were no great leaps forward. Instead, the decades leading up to Deep Blue were all about steady improvement. The algorithms behind chess programs were made more efficient, but more importantly computer hardware advanced considerably. And as processing speeds increased so did the ability of chess programs to look further ahead in games.

The amount of computer memory available also grew rapidly. In the 1950s memory was in short supply and as such of little use in chess programs, but as memory became cheaper and more plentiful, new options opened up. Chess programs could now remember the scores they assigned to moves so that they no longer had to waste time repeating the same analysis over and over. Memory also allowed computers to draw on databases filled with centuries of human chess knowledge, from how to counter particular attacks to how best to end a game.

In artificial intelligence parlance this was a "brute force" approach, where processor speeds and memory banks rather than clever software tricks drove the improvements. By the end of the 1970s these developments had allowed chess programs to steadily narrow the gap between them and human players. The first home computers could now run chess programs that could challenge amateur players, and in 1978 Northwestern University's Chess 4.7 computer even managed a draw against British international master David Levy.

Add another twenty years of software refinement and rapid hardware advances, and the result is Deep Blue, a chess computer capable of assessing 200 million positions a second and defeating world chess champions.

But even as Deep Blue vindicated the half century of artificial intelligence research spent on chess, those working in the field knew that a far more daunting challenge now lay ahead: teaching a computer to master the ancient Chinese game of wéiqí, a game best known in the West by its Japanese name, go.

The origin of wéiqí, which is pronounced "way chee," is unknown. The game is thought to be around three thousand years old and there are many Chinese legends about its creation. One concerns the legendary and revered Emperor Yao, who is said to have ruled China for a hundred years and gave the nation the calendar. The tale says Yao invented wéiqí in an attempt to change the ways of his eldest son Danzhu, an unruly playboy who viciously oppressed ordinary people. Danzhu is said to have mastered the game, but it didn't change his ways. Unable to reform his son, Yao banished Danzhu from the kingdom and appointed a farmer as China's next emperor.

Regardless of the game's origin, wéiqí became an important part of Chinese civilization. Books on how to play it were written and poets penned odes to the game. By medieval times wéiqí had become one of the four arts that Chinese civil servants were expected to know alongside calligraphy, painting, and how to play the gǔqín, a string instrument.

By then wéiqí had taken on the form that we know today, with its square board divided into a grid by nineteen horizontal and nineteen vertical lines, rather than the seventeen by seventeen grid of old.

The game's core rules are deceptively simple. One player has a pot of white stones that are his or her pieces, the other plays with black stones. Taking turns they place one of these lens-shaped stones on any one of the three hundred and sixty-one intersections of the board's horizontal and vertical lines. No stone can be placed on an occupied intersection and stones do not move once placed on the board.

The goal is to capture enemy stones and control territory. Players capture enemy stones by surrounding them with their own stones until there are no empty intersections next to their opponent's pieces. To control territory players must surround an area of the board with their stones so that all lines leading out of that area are blocked and no enemy stones are within that area. When no more legal moves are possible or both players agree to stop, the game ends. Players then add up how many enemy stones they took prisoner and the number of empty intersections in their territories. The winner is the player with the highest score.

Although simple in its rules, the size of the board coupled with the intricacies of capturing and recapturing territory and stones create a game of great complexity, closer in spirit to an entire military campaign filled with local battles rather than the single battle represented in chess.

While the Chinese began the process of formally trying to master the game's complexities, it was Japan that took the science of playing wéiqí to the next level. The game reached Japan around AD 500 and, under its new name go, it became popular among aristocrats. By the eleventh century go was played by aristocrats and commoners alike.

Go's ascendance in Japan didn't stop there. In 1588 Toyotomi Hideyoshi, the samurai and go fan who secured most of Japan under his rule earlier that decade, sponsored the first national tournament and the creation of a system for ranking the best players. His successor Tokugawa Ieyasu, who completed the task of unifying Japan, gave the winner of that first tournament, Honinbo Sansa, a plum job as the man in charge of all matters related to go. Sansa used his position to open government-funded academies dedicated to training future go masters.

But when the shogunate collapsed in 1868, the go academies went down with it. The government money vanished and go schools struggled to attract pupils willing to pay the fees they now had to charge. Pupils were reduced to wandering the streets and trying to eke out a living by playing games for cash. Go's decline wasn't just limited to the academies; the whole country lost interest. The opening up of Japan after more than two centuries of self-imposed isolation led to an influx of new foreign goods that drew the Japanese away from their traditional interests.

But go's spell in the doldrums proved short lived. By the early twentieth century, interest in the game had revived thanks to the rise of the country's new middle class. Japan's middle classes embraced the game because of its parallels with Taoist philosophy and the belief that it harbored valuable lessons for business, such as the need to focus on long-term goals rather than short-term battles.

By the middle of the 1920s the revival had led to the formation of the professional go association Nihon Ki-in, which reformed the rankings system, opened annual tournaments to the public, and imposed time limits to stop professional games from dragging on for months as they often did in the old days.

Although popular in Japan and China, go struggled to find a wide audience outside East Asia. Even now the game remains little known in the West. But go did find a keen fan base among Western computer scientists, mathematicians, and engineers.

Despite its following among the technically minded, few in artificial intelligence circles tried developing go-playing programs, and even in the early 1980s researchers still believed it would be many decades before go masters would need to worry about digital competition.

As Hans Berliner, organizer of the World Computer Chess Competition, told the BBC in 1982, when asked what stage artificial intelligence would be at when a computer could beat a go grandmaster: "For one thing very close to where people would get very paranoid because it would have gone a very long way. They will be at a point where they could start replacing ordinary humans in all sorts of every day environments. I think that when we get to that point, there will be very few things that we won't be able to do."

The pessimism about computer go lay in the game's complexity. For any given situation in chess there are, on average, thirty-five possible next moves. In go there are two hundred and fifty.

As if that wasn't enough there was an even thornier question for computer go: how to make a meaningful judgment about whether a move is any good. It's a question so slippery that even the best go players can find it challenging to predict who will win a top-level contest mid-game.

"In chess we have two armies that face each other and whoever has got the strongest army is usually in ascent, so it's not so difficult to estimate who is ahead and who is behind," says French computer go researcher Rémi Coulom. "In go it is a lot more difficult because you cannot count the number of stones on the board and say: 'Oh, I am one stone ahead.' It does not work like this. The evaluation of who is ahead or not is a lot more subtle. It's very dynamic. You have stones but it is difficult to determine whether they will live or die or which territory will belong to which player."

So even by the time Deep Blue wounded Kasparov's pride, those creating computer go programs were so pessimistic about their efforts that one said it could take another century before they would achieve a similar victory. Even the team that made Deep Blue felt daunted by go, admitting that they would struggle to make a machine that could challenge a mediocre go player anytime soon.

Nonetheless with Deep Blue's success in the bag, go emerged as the new holy grail for artificial intelligence research. Coulom was among those who made the switch from computer chess to go. He had started making computer chess programs as an eighteen-year-old in the early 1990s. He called his program "Crazy Bishop," a play on how the French word for crazy and fool—France's equivalent of the bishop—are the same: *fou*. Coulom worked on Crazy Bishop for years, making regular trips to the annual Computer Olympiad where game-playing computer programs battle in machine-versus-machine contests.

Then, in 2004, he became an associate professor at the University of Lille 3 and found himself overseeing a master's degree student who wanted to work on computer go. Coulom was only dimly aware of the game so he asked fellow French artificial intelligence researcher Bruno

Bouzy, who he had met at a Computer Olympiad, to help supervise the student. Bouzy was already involved in the computer go scene and told Coulom how he and others were using the Monte Carlo method to help computers play the game.

The Monte Carlo method originated in the Los Alamos National Laboratory in New Mexico as its scientists raced to create the atomic bomb during the Second World War. The Manhattan Project team wanted to assess how far neutrons from a nuclear explosion would penetrate different materials, but with so many variables and materials to assess, doing all the calculations would have taken a prohibitively long time.

Their solution was the Monte Carlo method, which got its name from the casino district of Monaco. Instead of trying to calculate every possibility, the Monte Carlo method randomly sampled fewer possibilities to generate an answer that, while falling short of being the perfect answer, was a good enough prediction of what might happen.

It's not unlike a political opinion poll. Instead of asking every person in the country who they will vote for, pollsters ask a large enough group of people to get an answer that is pretty close to the one they would get if they asked everyone.

Bouzy and others had adopted the Monte Carlo method as a way of evaluating which of the moves available to the computer would likely be the best. It helped make go programs better, but the improvement was minor. Then Coulom hit on the idea of blending the Monte Carlo method with the minimax process used in computer chess to build game trees of all the possible moves and where they could lead. He called his approach Monte Carlo tree search.

Instead of creating a game tree of every possible move, as is the case with minimax, Monte Carlo tree search plays out entire games fast, using random moves, and notes whether these games result in a win or a loss. As the program plays more and more random games, it starts homing in on which moves and branches of the game tree look most likely to result in a win. While the approach would mean the program might miss the very best move because it didn't happen to play out that game at random, the results were good enough and, after all, human opponents aren't always perfect either.

"It took a few months before I made it really work because it takes time to program and test but the progress was rather fast," recalls Coulom. "When I invented the idea I was really excited about it. I didn't know how it would turn out but I had the intuition that it was a big idea."

The idea worked and when Coulom took the resulting program, which he named Crazy Stone, to the 2006 Computer Olympiad in Turin, Italy, he walked away with a gold medal. Monte Carlo tree search was the biggest breakthrough in computer go for years and soon he and others were refining and enhancing the process to deliver better and better programs.

Now, instead of having computer go programs that would barely challenge a novice, they could mount a decent challenge to or even beat experienced players. Soon programs such as Zen, MoGo, and Crazy Stone began clocking up victories against professional go players, including Crazy Stone's victory against the top-ranked player Norimoto Yoda at a computer go tournament in Tokyo.

But these were no Deep Blue–style victories, for Crazy Stone and its peers had help: a four-stone handicap. Go's handicap system, first developed more than three hundred years ago in Japan's shogun-sponsored academies, is designed to balance the game so that weak and strong players can play together.

Depending on the gap in skill between the players, the stronger player incurs a handicap where the weaker player gets to put down a set number of stones on the board before their superior opponent can move. And a four-stone handicap is huge, an advantage considered big enough to give a strong amateur a chance of defeating even the world's highest-ranked players.

Even so, Yoda, who holds the highest possible rank in go, wasn't pleased about being defeated by a machine. "He was not happy to lose. He's a professional player and prefers to win, but he is a professional player and Japanese people have very polite manners so he didn't show much anger," says Coulom. "But a four-stone handicap doesn't scare him very much."

While Crazy Stone did not have professional go players quaking in their boots, its abilities were more than a match for its creator. "To be honest I'm not a strong go player myself and my program has reached a level where it is much stronger than I am," admits Coulom. "So when I watch a game, most of the time I don't understand what is going on."

Monte Carlo tree search was an important leap forward but if computers were to take on go masters properly something more was needed. And researchers soon homed in on another promising approach: deep neural networks.

Neural networks seek to mimic the tangle of neurons in the human brain to create programs that learn how to do complex things through a process of trial and error. The idea originated in the 1950s, but the limits of computing power back then meant that even the biggest neural networks were limited to just a few hundred "neurons." If an animal neuron was equal to a neural network neuron, these early networks would lie somewhere just below roundworms on neuron count and be dwarfed by the towering intellect of a jellyfish.

But, as Deep Blue proved, in computing, time and hardware advances can overcome a lot and by the 2010s neural networks had become a viable option for artificial intelligence research.

Neural networks for computer go soon began producing promising results in computer go. In late 2015 Facebook revealed it had been using neural networks combined with Monte Carlo tree search to train computers to play go and that the resulting programs played the game slightly better and also in a more humanlike way.

Impressive as Facebook's go programs were, they were still incapable of defeating top players. It was progress but the Deep Blue of go still seemed years away.

But then in January 2016 Google dropped a bombshell. The company's London-based artificial intelligence research subsidiary DeepMind, which it bought for $600 million in 2014, had created a program called AlphaGo that had defeated European go champion Fen Hui without a handicap.

AlphaGo was a hybrid. Part neural network, part Monte Carlo tree search. The program had been fed on a diet of thirty million professional go matches and then played the game against itself countless times until it had, through trial and error, learned to play with the skill worthy of a professional player.

The news shocked artificial intelligence researchers and go players alike. Go experts who watched its game against Hui admitted that they would never have thought one of the players was a machine. No one had expected a computer to play go this well for at least another decade. After getting over the initial shock, go experts pointed out that while Hui was the best player in Europe he was "merely" the world's 633rd best player.

But Google already had answer to this: a showdown between AlphaGo and Korean go master Lee Sedol, the world's most accomplished player of the game, that would take place in March 2016. Just like Kasparov's battle with IBM before it, the five-game match in Seoul captured the world's attention. Once more man would face machine.

Before the event, Lee—like Kasparov before him—was confident. "I heard Google DeepMind's AI is surprisingly strong and getting stronger, but I am confident that I can win at least this time," he told the *Financial Times*.

But the first game, held on Wednesday March 9, 2016, ended not in a victory for humanity's champion but for the digital challenger. After the match, Lee admitted to journalists that he never thought he would lose.

AlphaGo then won the second game, leaving it just one victory away from winning the entire contest. And when Lee took on AlphaGo for a third time on March 12, 2016, the computer program claimed that its third consecutive victory and won the match. Lee politely accepted his defeat, admitting that the pressure had got to him, but agreed to play the last two games all the same.

The next game finally saw Lee beat AlphaGo after he forced the machine into making a mistake that it could never recover from. Lee's victory prompted cheers and applause from the audience. "I've never been congratulated so much just because I won one game," he said. His victory was too little, too late, but the reception that greeted it suggested that deep down everyone wanted man, not machine, to win.

But humanity's comeback was brief. In the final game AlphaGo defeated the go master yet again. In the wake of the program's success, Korea's national go association awarded AlphaGo an honorary professional 9-dan title, the highest possible rank in the game, and the website Go Ratings named DeepMind's AI the world's fourth-best player.

Lee's defeat spawned widespread debate about artificial intelligence and what it meant for our future for unlike Deep Blue, AlphaGo was more than a publicity coup for Google.

While Deep Blue's dedicated chess-playing hardware had little wider application, Google has big plans for the self-teaching technology DeepMind had used to end humanity's dominance in go. Demis Hassabis, the founder of DeepMind, told reporters that the lessons from AlphaGo could enable the creation of personal assistants for our smartphones that would make Apple's Siri look like Clippy, the much maligned paperclip assistant from late 1990s editions of Microsoft Office.

Other potential applications could include controlling self-driving cars, vastly improved computer translations, and even diagnosing diseases and picking the best treatments. And, probably of greatest importance for Google's shareholders, the same technology could also transform online advertising. Instead of our web browsers being filled with ads that follow us around the net like creepy stalkers based on sites we've already visited, AlphaGo-style AIs could anticipate what we are looking for before we even search for it.

While these applications may be some way off, Google is already using DeepMind's technology to do more than play games. A few months after AlphaGo's victory over Lee, Google revealed that DeepMind's game-playing tech was now reducing the electric bill at its data centers by manipulating servers, fans, windows, and cooling systems in new ways. In just a few months, Google said, the program had shaved several percentage points off the company's power bill. That's a hefty saving given that in 2014 Google's electricity use was comparable to that of every household in a city the size of Austin, Texas.

AlphaGo has brought Turing and Shannon's dream of an artificial intelligence much closer to becoming reality and for some that's a frightening prospect. What will people do if computers can do their jobs better and cheaper than they can? Even some in Silicon Valley worry, including PayPal and SpaceX founder Elon Musk, who has likened artificial intelligence to "summoning the demon." Physicist Stephen Hawking, meanwhile, has warned that artificial intelligence "could spell the end of the human race." Even DeepMind's Hassabis has said that artificial intelligence could do harm if misused.

Just as Berliner predicted in 1982, now that computers can beat go masters, fears about artificial intelligence are rising. That no one understands how AlphaGo achieved its victories won't ease those concerns either. The "thought" processes of neural networks are opaque, there is no computer code to read or formulas that explain it. Even DeepMind doesn't know how AlphaGo plays go.

For now that doesn't matter. So what if we don't know why a computer is better at go than the finest human players? But whether we will feel the same when the artificial intelligences that will follow in AlphaGo's footsteps are driving our cars and making life and death decisions about our healthcare is another question entirely.

14

# TRIVIAL PURSUIT: ADULTS AT PLAY

## *How Trivial Pursuit broke games out of the toy box*

The afternoon air bit as Chris Haney walked through the streets of Montréal. It was an icy twenty-one degrees Fahrenheit and the overcast sky a mournful gray.

The day hadn't really gone according to plan for the *Montreal Gazette* photo editor. He should have been at home, in the warm, drinking beer and playing Scrabble with his friend Scott Abbott, a sportswriter for the Canadian Press news agency. But they couldn't find the Scrabble set, so here Haney was, trudging to the store to save the day.

On returning home Haney removed his coat and, as they cracked open the beers, he complained to Abbott about the game's seven-Canadian-dollar price tag, equivalent to just over sixteen U.S. dollars in today's money. Abbott nodded in sympathy, having bought several replacement Scrabble sets himself over the years. The people behind Scrabble must have made so much money.

By the time their first game of Scrabble was over, the pair had reached the same conclusion: we should make a game—we would make a fortune.

For 99 percent of people who reach a similar conclusion, the story usually ends with a never-finished game or one that no one wants to publish. At best they might get their game published, only for it never to sell enough to pay even a dime in royalties. But the game Abbott and Haney conceived that dismal afternoon really did make them, and several other people, a fortune. Its name was Trivial Pursuit.

Their kitchen-table discussion of what kind of game to make was brief. Neither wanted to make a word game like Scrabble or another Monopoly knock-off. Abbott suggested that they make a trivia quiz game because no one had made a good one yet.

Forty-five minutes later they had a prototype. They sketched out the board on a sheet of paper, creating a track that looked like a ship's wheel, before gluing it onto a spare square of cardboard they found lying around. For playing pieces they raided the other board games in Haney's collection. Finally, they wrote a bunch of trivia questions, organized into six categories: arts and literature; entertainment; geography; history; science and nature; and, finally, sports and leisure.

The players would travel the board landing on squares representing the various categories until they had correctly answered a question from all six categories. After that players would race back to the center of the wheel and have to answer one final question in a category chosen by the other players to win.

Abbott and Haney envisaged their game as something for adults rather than the children and parents who accounted for most game players at the time. So, as they expanded the game with the help of various trivia books, they selected questions that they felt their generation would enjoy answering or at least learning the answer to.

Questions of largely inconsequential knowledge from "What is the first flavor in Life Savers candy?" and "How long did Yuri Gagarin spend in space?" to "Where are the Nazca Lines?" and "Who told his hotel staff not to mention the war?"

These questions were perfect fodder for the baby boomer generation, says Philip E. Orbanes, who was a senior vice-president at Parker Brothers in 1979. "The baby boomer generation grew up with TV and had a lot of fond memories of early TV shows and movies and trivia in general," he says. "It was much more of an instinctive topic for people of my generation. We had been challenging each other with trivia questions for years and years. Do you remember this show? Who was the actor who starred in it? And what was the color of the lace dress worn by this member of the royal family on this occasion?"

Questions added, the pair christened their game Trivia Pursuit but changed the name after Haney's wife said that Trivial Pursuit sounded better.

The pair then began thinking about how to get Trivial Pursuit into stores. One thing was clear: they both lacked the money they would need to do it themselves. So they invited Haney's brother John, a former ice hockey player, and attorney Ed Warner to join them in trying to turn their trivia game into a business.

But Abbott and Haney's vision for a high-quality game featuring a large folding board and a huge stack of trivia question-and-answer cards was still too expensive for them to manufacture. So the business partners began asking everyone they met to give them money in return for shares in the business. They also used the offer of shares to persuade the people they needed to help them turn the game into a finished product to work for free, among them Michael Wurstlin, the graphic designer who gave the game its distinctive, almost Victorian, look.

After two years of slog the company had attracted thirty-two investors and raised enough money to publish the game, but only just. The money allowed them to produce one thousand copies of the game, but this was way too few to take advantage of any economies of scale, so each copy of Trivial Pursuit cost seventy-five Canadian dollars to produce.

When they revealed the price to stores they hoped would stock the game, jaws dropped to the floor with a thud. Even the few stores that agreed to buy the game refused to pay more than twenty dollars a copy since that would still mean a retail price of forty dollars, more than five times the price of Scrabble.

What made Trivial Pursuit's eye-watering price tag seem all the more outrageous was that Abbott and Haney were trying to sell it to stores at the peak of video-game mania. North America had gone nuts for video games following the release of Space Invaders in 1978. Arcade halls were opening everywhere and the constant noise of Pac-Man, Asteroids, and Donkey Kong machines had become the soundtrack to daily life. Coin-operated video games were invading supermarkets, infiltrating airports, appearing in golf clubs, and even taking root in dentist waiting rooms. Atari game consoles now topped children's Christmas lists.

Selling an expensive board game with Victorian artwork in this age of digital thrills seemed as out of step with the times as trying to get people to swap their Sony Walkmans for gramophones.

Things didn't get better when the pair took their game to the Toronto and New York toy fairs in early 1982. The board game industry's verdict was unanimous. Video games were the future, not this overpriced flashback. The pair returned to Montréal with barely enough orders to cover the cost of attending the events and their hopes crushed.

Haney took it hardest. Getting Trivial Pursuit to this stage had wiped out his savings and he had sold most of his photography equipment to keep the dream alive. The sole consolation was that he had persuaded his mother not to invest in the game out of fear that she would lose all her money.

But in summer 1982 the Canadian bookstores and toy retailers that took a gamble on Trivial Pursuit back in November 1981 began reordering the game, reviving the pair's confidence in their creation. Abbott borrowed forty thousand dollars from his father and the company used the money to secure seventy-five thousand dollars' credit from a bank. They printed another twenty thousand copies of Trivial Pursuit, enough to make selling it profitable. By the end of 1982 almost all of the new copies had been sold.

From there Trivial Pursuit's rise was unstoppable. In 1983 Selchow & Righter, the New York company behind Scrabble and Parcheesi, bought the rights to publish the game in the United States and began promoting it by sending copies to the celebrities mentioned in the game's stockpile of six thousand questions.

The celebrities returned the favor. Trivial Pursuit won public endorsements from stars like Charlton Heston and Frank Sinatra, and appeared in *Time* magazine after the cast of *The Big Chill* got hooked on the game.

Selchow & Righter soon found itself straining to meet demand. The company's other new products went ignored as the family-run firm's three hundred employees worked flat-out to keep Trivial Pursuit on the shelves.

Trivial Pursuit proved that those who believed video games would make board games history were deluded. In 1984 the video game business went into free fall, toppled by its own overconfidence and the rise of videocassette recorders.

Trivial Pursuit, meanwhile, became one of the biggest sensations of the eighties. By the end of 1984, twenty million copies had been sold, but even sales that high couldn't satisfy the public's hunger. Trivial Pursuit was so sought after and so hard to get hold of that secondhand copies were soon changing hands for as much as sixty dollars.

The board game business had never seen anything like it. "In a really good year in the late 1970s, early 1980s, Monopoly might sell three million copies," says Orbanes. "Trivial Pursuit sold twenty million in the first year and it was thirty-five dollars at retail."

Mike Gray, a game designer for Milton Bradley at the time, remembers how desperate people were to get the game. "I had neighbors that I didn't know who would hear that I worked at Milton Bradley and would come and knock on my door and say do you have Trivial Pursuit?" he recalls. "Of course I did and they wanted to borrow it or they wanted me to come down and play with them. You couldn't get it anywhere. Who would have thought that a forty-dollar question-and-answer game would be the game everybody had to have?"

Image was central to the game's success. Trivial Pursuit wasn't just popular; it was a game to be seen with. Trivial Pursuit's sophisticated look, production values, and high price all reinforced the perception that this was a game that adults could play without shame, a game you wouldn't be embarrassed to suggest playing at a cocktail party.

The nostalgic appeal of the questions was equally important. Trivial Pursuit reconnected the baby boomers, who were now saddled with mortgages, kids, careers, and pension plans with the television shows, music, and defining moments of their fast-fading youth.

Trivial Pursuit gave the decade's thirtysomethings a chance to indulge themselves, to measure the quality of one another's middlebrow knowledge, and reassure each other that a head filled with three decades of random pop culture tidbits had value. As avid Trivial Pursuit player Barbara Handler told *The New York Times* in 1984: "It's like a sickness. You want to see how much garbage you know."

Sociologists speculated that the Trivial Pursuit phenomenon was a symptom of "cultural exhaustion" driven by young adults who wanted to revisit their childhood and hide from an uncertain world.

Another factor was leisure. Since 1960 the amount of time Americans spent at work had dropped by two hundred hours a year and the growth of supermarkets and labor-saving domestic appliances had slashed the amount of nonwork time swallowed up with chores. The baby boomer adults of eighties America were enjoying a bounty of leisure time that no generation before had.

The rapid expansion of leisure time prompted much hand-wringing. Commentators worried about "leisure shock" or "leisure overload" as a generation found themselves with so much spare time that they had little idea what to do with it but slump in front of the television. Soon the self-help sections of bookstores were stocked with titles like *Recreation for Today's Society* and *Free Time: Making Your Leisure Count* that offered people advice on making the best use of their surfeit of free time.

While this growth in leisure time came to a swift end after 1985 due to rising prices and job insecurity, Abbott and Haney's game had launched at precisely the right time to reap the rewards of the baby boomers' relatively bountiful leisure time. Trivial Pursuit became the must-have game of a generation that declared that adults were no longer content to sacrifice play on the altar on maturity.

Whatever Trivial Pursuit had tapped into, it was big. People lapped up the main editions, gobbled up specialized versions focused on movie or sports trivia, and, when those were exhausted, splashed out on card packs that replenished the supply of questions.

By March 1986 an estimated one in four U.S. homes had a copy of the game, around the same as the number of households that owned a microwave oven. Everyone who took shares in Trivial Pursuit walked away with hefty returns while the company's founders got to spend their millions on residences in the Bahamas, private jets, racehorses, ice hockey teams, and golf courses.

But they had done more than get rich. They had also proved that adults still wanted to play games—just not the same ones as their children. "Trivial Pursuit showed us that there was an adult game market and I think that's really significant," says Gray.

Soon every board game publisher was on the hunt for the next Trivial Pursuit and it was Milton Bradley that found it. The game was called A Question of Scruples, but unlike Trivial Pursuit it was designed not to celebrate the trivia-soaked brains of baby boomers but to probe their morals.

A Question of Scruples' creator Henry Makow was also Canadian; he was an English literature lecturer at the University of Manitoba and an aspiring writer.

In 1983 he began working on an article about what he saw as the moral hypocrisy of his generation. "The baby boomers felt it was all right to rip-off the government, yet at the same time they thought they were morally superior," he says. Baby boomers might talk the talk on fashionable issues like protecting the environment or world peace, he felt, but when it came to the small stuff, the ethical challenges of daily life, they lacked scruples.

To gather evidence for his argument, he invited several friends and relatives to his home for an evening of moral debate, and to get the discussion going Makow handed his guests a questionnaire with a list of ethical dilemmas. Would they pose nude in a magazine for ten thousand dollars or admit they had smoked marijuana if asked by their kid? Would they remind a waitress if she forgot to charge them for drinks or inform a friend that his fiancée is coming onto him?

The evening was meant to help Makow make his point but it was more fun than anyone expected, and when one of the guests said it would make a great game Makow ran with the idea. The article never got written. Instead Makow took his questionnaire and spent the first six months of 1984 turning it into a game.

He expanded the range of conundrums and gave players the ability to challenge each other if they felt the person answering the dilemma on the card wasn't telling the truth. He came to see his game as almost an antidote to Trivial Pursuit: "I felt Trivial Pursuit was trivial. I wanted something substantial. I also wanted people to take a seminar in everyday morality and enjoy it."

Following in the footsteps of Trivial Pursuit's creators, Makow published the game himself with the help of a thirty-five-thousand-dollar windfall he got from a real estate investment made by his mother. And in the wake of Trivial Pursuit's success, retailers were much more willing to stock the game.

Soon Makow's game was on sale in Sears Canada department stores and Birks gift shops. In its first four months on sale, ten thousand copies of A Question of Scruples were sold and Makow found himself in talks with Milton Bradley about taking his game beyond Canada.

While Makow accepted Milton Bradley's offer, he and the game publisher didn't see eye to eye on the game, especially when the Massachusetts game giant decided it wanted to change many of the questions. "They added questions I thought were lame, I thought many of theirs weren't moral dilemmas," says Makow. "The best moral dilemma involves a universal sense of right and wrong, and you're not sure which is which."

While most of Milton Bradley's additions stayed, Makow vetoed some. "One stands out: 'You are driving and need to masturbate . . .

do you?'" he recalls. "I was so outraged when I saw this I faxed it to the president of the company. The execs were chewed out."

While many of the moral challenges that did make it into the final game still ring true today, others make A Question of Scruples read like a time capsule of eighties attitudes. "Your teenage daughter is dating a boy of another color. Do you encourage her to date boys of her own race?" asked one card. Other dilemmas included: "You are smoking at a meeting. Someone is coughing and showing discomfort. Do you finish your cigarette?"; "A neighbor is beating up his wife. Do you call the police?"; and "Your doctor makes sexual advances to you. Do you report him/her to the medical association?"

Makow's game of moral quandaries became a huge seller. As 1986 began A Question of Scruples was selling faster than Trivial Pursuit and 1,500,000 copies had been sold. By the end of 1990 Makow's game had found its way into seven million homes across the world and, like Abbott and Haney, Makow found himself a rich man. "I'm afraid I did not handle the sudden wealth well," he says. "My lifestyle remained the same except I had to devote too much time and energy to managing my money."

After making another, this time unsuccessful, game, Makow retreated to his writing and later reinvented himself as a devotee of outlandish conspiracy theories that range from The Beatles being Satanic Illuminati puppets to the destruction of the World Trade Center being an "inside job."

Others also tried to tap into the zeitgeist by marrying the newfound appeal of games for adults with the eighties obsession with getting rich. First out of the blocks was 1985's the Yuppie Game, a celebration of young urban professionals—the baby boomers who were getting rich on the back of a booming stock market.

The game was bashed together in a couple of hours by two friends after a night of Trivial Pursuit, wine, and brie. The goal was to amass the trappings of the yuppie success from BMW 320i cars and high-rise condos to purebred Himalayan cats and etched goblets to quaff wine from.

The Yuppie Game's greed-is-good theme, however, was tame in comparison to 1989's Trump: The Game. Based on multimillionaire businessman and future U.S. president Donald Trump's bestselling book *The Art of the Deal,* Trump: The Game had players speculating and trading real estate in the hope of walking away with hundreds of millions of dollars in profit.

The idea of turning Trump's book into a game came from the Chicago toy invention agency Big Monster Toys, one of the companies founded by the former alumni of Marvin Glass and Associates. After completing the prototype, codesigner Jeffrey Breslow secured a meeting to pitch the idea to the businessman.

"At that time Trump was a germaphobe," says Breslow. "We were told: 'Don't even stick your hand out, he's not going to shake your hand.' We had a meeting at two o'clock and at two o'clock on the money we were in his office—they didn't keep us waiting a second. The meeting was hello, blah, blah, blah, let's see what you have. It wasn't small talk or anything else, it was boom, boom, boom.

"So I set the game up on the conference room table and rolled the dice and he said: 'I like it, what's next?' I mean we played the game for ten seconds, it wasn't like he was going to sit down and play a board game even though it was his game. So he saw it and said: 'Ok, what's next?' So I said: 'Well, er, I'm gonna pitch it to some of my biggest clients and come back to you with a deal.' We were out of the office in ten, fifteen minutes."

After getting Milton Bradley to buy the rights, Breslow returned to give the news to Trump. "Right on the nose I was in his office and it was: 'Hello, how are you? What have you got?'" says Breslow. "It wasn't 'How's the wife and kids?' Everything about him was very business.

"So I made the deal with him and he said: 'What's next?' I said: 'Well, it would be really important for you to come to a toy fair in New York when we're pitching the game.' And he said: 'I'll be there. Anything else?' I said: 'Well it would be really neat if you could come up to Milton Bradley in East Longmeadow where the games are coming off the assembly line for publicity.' He says: 'I'll be there.'

"So he flew his helicopter up there and came to the toy fair. Everything he said he was going to do he did. It was an extremely successful game, it did extraordinary numbers. It was a big hit at the time."

While Trump's game did big numbers, its sales paled next to Pictionary, the game that had replaced Trivial Pursuit and A Question of Scruples as the hottest game on the block by 1989.

Pictionary's origins date back to 1981 when its creator Rob Angel was sharing a house in Spokane, Washington, with a few high school friends. The housemates had gotten into the habit of playing a version of charades where they plucked a random word from a dictionary and then had to half-draw, half-act out the word.

Angel began trying to turn the game into something more structured and wrote down a number of ideas on a legal pad, but as the demands of work took over, he forgot all about the project. Then in 1984 he relocated to Seattle and while unpacking rediscovered his old notes and started thinking about the game again.

The idea of drawing charades seemed like a good one, but he couldn't imagine people playing a game using a dictionary and had no idea how to get around that. But then he played Trivial Pursuit for the first time and realized the answer was to put the words players had to draw onto cards they would collect while moving around a board.

Encouraged by Trivial Pursuit's success, he teamed up with graphic designer Gary Everson and accountant Terry Langston to hone the game and put it into production. Using money Angel borrowed from his aunt and uncle, the trio paid for one thousand copies to be printed and then spent one pizza-and beer-fueled week assembling each set at his home from the mountains of boxes, cards, and playing pieces they had ordered.

On June 1, 1985, Pictionary launched and, just like Trivial Pursuit, its understated white-text-on-black box art and above-average price positioned the game as aimed at adults not kids. Sales were sluggish until the trio began going to the stores where Pictionary was being sold and demonstrating the game to the public. They quickly found themselves hosting mass games of Pictionary in front of dozens of shoppers and handing out box after box for people to take to the checkout.

Word of this new game spread fast. Soon the original thousand copies were gone. So the trio borrowed more money and printed another ten thousand sets. By Christmas every single copy had been sold.

By summer 1986 they had sold another forty-five thousand Pictionary sets and were struggling to keep up with the demand, so they began looking for a publisher. They tried Worlds of Wonder but walked away after they were offered disappointing terms. Milton Bradley made a promising offer but the deal collapsed when they looked at the small print. So the game ended up in the unlikely hands of Western Publishing, ironically a children's book publisher that had never released a game before.

Western Publishing's inexperience didn't matter. By the end of the year more than three million Pictionary sets had been sold and by the end of 1988 the game had eclipsed A Question of Scruples, with sales passing the nine million mark in North America and retailers like Kmart requesting fifty thousand copies a month. The game went on to sell more than thirty million copies worldwide.

By the start of the nineties store shelves were groaning under the weight of games inspired by Trivial Pursuit, from the word-guessing games Taboo and Scattergories to TriBond, which challenged players to identify the connection between three words while moving around a triangular board.

Board game collections had become a fixture at adult parties. Trivial Pursuit might have been one of the greatest fads of the 1980s but it permanently altered playing habits. Before the Canadian trivia game stormed the world, board games were viewed as kids' stuff. Sure, some adults played war games but not many. Only games with an intellectual image, like chess or Scrabble, or strong gambling associations, like backgammon, seemed immune to the charge that playing was childish.

But with its appeal to trend-conscious yet nostalgic young adults, Trivial Pursuit broke board games free of the toy box for good.

15

# PANDEMICS AND TERROR: DISSECTING GEOPOLITICS ON CARDBOARD

*What board games teach us about disease, geopolitics, and the War on Terror*

By February China was worried.

Since November hundreds of people in the southeastern province of Guangdong had been struck down by a mysterious fever that rapidly developed into breathing problems so severe they needed to be hospitalized. The illness had killed five and the rate of infection was accelerating.

At first the Chinese health authorities covered up the sickness. They gagged the media and said nothing about it to other countries. But now it was getting out of hand. So on February 11, 2003, China finally informed the World Health Organization.

Ten days after China alerted the world, Liu Jianlun, a doctor from Guangzhou, the capital of Guangdong, checked into the Metropole Hotel in Hong Kong. He had felt unwell for days but the symptoms were not too bad so he shrugged it off and pressed ahead with his plan to attend a family wedding.

When Jianlun woke the next morning in his hotel room, however, the fever had worsened and he found himself struggling to breathe. Coughing and straining for air Jianlun rushed to the nearby Kwong Wah Hospital and was taken straight into intensive care. He died there a few days later.

Shortly afterward others who had stayed at the hotel while he was there began coming down with the same illness. Among them was a businessman who had stayed in the room across the hall from Jianlun and flown to Hanoi, Vietnam, before the sickness set in. Soon people were falling ill with the same disease in Beijing, Singapore, Taipei, and Toronto.

By then the World Health Organization had formed an international research team to try and identify the cause of the disease they were now calling severe acute respiratory syndrome, or SARS for short.

As the outbreak intensified Hong Kong descended into panic. Public transport, restaurants, and shopping malls became deserted. People who could work at home opted to stay away from the office.

At the end of March, the residents of apartment block E of Amoy Gardens, Hong Kong, awoke to find police and medics in protective clothing sealing them off from the outside world because two hundred residents had been infected.

The world watched on, wondering if this was the start of a terrifying global pandemic similar to the 1918 influenza outbreak that claimed the lives of at least fifty million people.

To everyone's relief that apocalyptic scenario never came to be. In April researchers isolated the SARS virus and linked it to the consumption of masked palm civets, a catlike creature whose meat was on sale in a market in Guangdong province. By the summer quarantine efforts had halted the spread of SARS, and by January 2004 the all-clear had been given.

The SARS outbreak infected several thousand people and killed more than seven hundred, but the rapid global response saved the world from an epidemic that could have been much, much worse.

Like millions of people the world over, Matt Leacock read the daily reports about SARS with a sense of dread. Hong Kong might be separated from his California home by ten thousand miles of ocean, but as the disease leapfrogged from city to city it was clear that only one infected person on one flight was needed to bring it to his neighborhood.

The outbreak unfolding on his television screen got Leacock

thinking about disease and how it was humanity's ultimate enemy, an invisible killer that knew no mercy. It was exactly the kind of enemy he was looking for.

Leacock was working on a cooperative board game where, instead of trying to defeat one another, players would unite against a common foe.

He had been making games since he was a kid but only a couple had ever been released. The first was a kingdom-building game called Borderlands. He made each copy by hand and then tried selling it without much success. "I don't know how many of those were sold but it was in the tens, so not very many," he says. "I think I still have some in the attic."

After that came Lunatix Loop, a slapstick demolition derby game featuring souped-up Trabants, the famously bad plastic-bodied cars built in communist East Germany. That game came and went largely unnoticed by the world too. Leacock didn't mind; he just liked making board games.

For his next project he found inspiration in The Lord of the Rings, a cooperative game based on the J. R. R. Tolkien novel. In the game players guided the hobbits through Middle Earth on their quest to throw the ring into the fires of Mount Doom while battling the forces of the dark lord Sauron, which were controlled by the game's cards and dice.

"I was fascinated by the fact that using just paper and cardboard you could create an opponent and make this little algorithm that is just run by cards sophisticated enough to get people really engrossed and talking," he says. "I also noticed that when I played it with the family we would feel good whether we won or lost. Competitive games with the family sometimes got a little troublesome because even if you won you felt bad, but with a cooperative game everybody felt good."

To make his cooperative game work, however, Leacock needed a common enemy for the players to fight and in SARS he saw that opponent. "Diseases felt like a natural enemy because they do almost reproduce in this algorithmic way, and I was also interested in things that could spiral out of control," he said.

The game he created, Pandemic, imagined a world threatened by four unnamed diseases that were spreading from city to city.

The level of infection in each city on the board was represented by the placement of disease cubes. After every turn city cards would be turned over and more disease cubes added to those metropolises. Once a city had three disease cubes, drawing its card would spark an outbreak and cause disease cubes to be added to every connected city. And if one of those cities already had three cubes another outbreak would start, sparking a chain reaction that caused the disease to erupt across the world.

Adding to the threat were the epidemic cards hidden in the deck. If drawn by a player these cards would increase the rate at which cities became infected and return the city cards already turned over to the top of the deck, so that places already infected were more likely to get an extra disease cube and spark an outbreak.

Leacock didn't base his system for how the disease spread on science, but with its chain reactions and intensification of outbreaks Pandemic captured the essence of how real epidemics can rampage around a globalized world. "I was far more concerned with creating a compelling game than I was in creating a simulation, but I was really happy to hear that people working in the field had played it, and were really engaged by it and not turned off by its representation of disease," says Leacock.

Pandemic's players, meanwhile, formed a multidisciplinary public health team not dissimilar to the one assembled to stop SARS. Each player had a specialty. They might be a scientist capable of developing vaccines faster than other players, a medic who could clear disease cubes more effectively, or an outbreak-containing quarantine specialist.

Using their different skills the players had to control and cure the four diseases before the outbreaks spiraled out of control and killed millions of people. This was, again, a simplification of the reality. Players didn't have to worry about war zones where conflict allows disease to fester, or religious fanatics killing health workers for trying to administer life-saving vaccines. Nor did they have to contend with a shortage of effective antibiotics caused by years of underinvestment in research and flagrant overuse of existing antibiotics.

Even so, saving the world from the diseases in Pandemic is far from easy. Since players can only do so much each turn and cannot be everywhere at once, the outcome of the game rests on a knife-edge. With every turn the world edges closer to an unstoppable pandemic and players must constantly weigh competing priorities. Should they clear the disease cubes from Sydney before an outbreak occurs or merely contain it? Is developing one of the vaccines needed to win the game this turn worth the risk of letting an outbreak happen in Johannesburg?

The game twists and turns like a roller coaster. One moment it might look like the situation is under control; the next, an epidemic card might leave players scrambling to reassert control.

As such Pandemic, despite its pared-down model, showcases how crucial teamwork and global coordination is to defeating diseases that threaten the lives of millions. The SARS outbreak was stopped because the nations of the world largely united and acted as one to defeat it.

But sometimes that doesn't happen. Even now, long after SARS demonstrated the value of international cooperation in fighting disease, bureaucracy and a lack of communication lets diseases get out of control. The Ebola outbreak in West Africa that began in late 2013 and claimed in excess of ten thousand lives is one such example. It took almost eight months for the outbreak to be declared a global emergency, a delay that made the task of getting Ebola under control significantly harder and probably led to many more deaths.

Pandemic, however, is an optimistic game where the world unites against disease. "No game is really neutral," says Leacock. "I'm trying to show the triumph of science instead of, say, violence. You actually use science in order to save the world, you're not just going around shooting people."

That point is emphasized on the game's box, which makes the

game's female scientist the focus of a squad of international infection-busters. "In the original cover that I got for the game, there was panic and there were burning bodies and mass hysteria," he recalls. "I was like, 'No, no, no! This is about a team overcoming, it's not about fear and horror.' That's part of why the brand is successful because it's optimistic and I think we need some of that in the world."

The world seems to agree. Since its debut in 2007 Pandemic has become one of the bestselling board games of recent times and spawned an array of expansions and spin-offs that have allowed Leacock to turn his hobby into a full-time occupation.

And with new disease threats from Middle East respiratory syndrome to the Zika virus flaring up on a regular basis, Pandemic's clash between man and microbe continues to feel relevant. "It's very topical, right? People are very concerned about this," says Leacock. "I think that's part of the reason why zombies are so popular these days in the media; people are just afraid of worldwide disease. That's helpful for the game I think because everybody can get behind fighting a disease. It's not often you find someone who is rooting for the disease."

While Pandemic's model of global disease management is simplified, it is a prime example of how board games can make the complexities of the world around us easier to comprehend. By bringing the subject of epidemiology to life on its board, Pandemic makes the concepts of how infectious diseases spread and what is takes to fight them intuitive to players in a way that a book or TV documentary could never do.

Pandemic is not the only game that, through its playful cardboard re-creation of reality, makes geopolitics easier to understand. Another example is the Cold War simulation Twilight Struggle, the flag bearer for a new wave of war games that are making war games easier to learn and quicker to play.

Twilight Struggle is the creation of Ananda Gupta and Jason Matthews, two war game fans who met in 2002 while carving out political careers in Washington, D.C. The inspiration for their game was Mark Herman's We the People, a 1994 war game about the American

Revolution. We the People breathed new life into war games by moving their complicated rules out of the rulebook and onto cards that players used during the game.

"We the People was extremely innovative because it had the card-driven system and it was very short—only an hour or two between experienced players—and yet it was clearly a historical war game," says Gupta. "Jason and myself saw it as the beginning of a renaissance."

That renaissance didn't pan out the way they expected. Instead of embracing the idea of war games that were easier and faster to play, the games that copied We the People's model simply layered on more and more complexity. So Gupta and Matthews decided to fight back with a game of their own—one that would put entertainment and speed before detail while remaining deep enough to satisfy war gamers.

And since Matthews had studied the Cold War, that long struggle between the capitalist West and communist East soon emerged as the obvious historical conflict to build their game around. "It was actually I who hit on the idea of the Cold War," says Gupta. "We were sitting in Jason's apartment and I was looking at his bookshelf and he had like fifty books on the Cold War and American foreign policy in the twentieth century, so I said how about the Cold War?"

Matthews wasn't sure at first, feeling that reducing the Cold War down to a two-player clash between the Soviets and the United States oversimplified what really happened. But after talking it over the pair realized that a two-player battle between the superpowers could be an excellent way to give players insight into how Washington and Moscow saw the world during those years of nuclear brinkmanship.

"I went into the project knowing a lot less about the Cold War than I do now," says Gupta. "I went in with a very Hollywood-ized view. I would think of John le Carré novels and spy movies and good and evil—all these very entertainment-driven perspectives—and I came out with a much more nuanced view."

Nonetheless the game does reduce the conflict to a clash of two nuclear superpowers. "If you put all the nuance of history inside a game then you can't spot the big themes," says Matthews. "We take the Cold War on its own premise that the only countries that matter are the United States and the Soviet Union. The reality of that is much, much more nuanced, but in order to illustrate that point to a public that may not even understand it or grasp it intuitively anymore, you have to kind of pound them over the head with it."

Core to the game are the cards the players collect and can deploy during their turns. Each card represents a crucial moment in the Cold War from the Cuban Revolution and President Nixon thawing relations with China to Soviet leader Mikhail Gorbachev introducing openness to the USSR with his Glasnost reforms.

These cards have their own unique implications for the game, changing the flow of the conflict, altering the influence of superpowers on different countries, and affecting their progress toward victory over their opponent. The Fidel Castro card ends U.S. influence over Cuba and hands it to the Soviet Union. The card representing Nixon's mission to China aids American activity in Asia. The Glasnost card lowers the risk of nuclear war.

But instead of having their own decks, players draw their event cards from the same pile, meaning that the U.S. player could be lumbered with a range of actions that aid the Soviets and vice versa. And when that is coupled by the push and pull of the jostling players' efforts to influence the world, Twilight Struggle makes being a Cold War superpower a constant round of crisis management where your hand is often forced by the swings and roundabouts of world events.

"The cards lead to this wonderful sense of paranoia or sense of crisis," says Gupta. "You draw your hand and suddenly you're not looking at this hand going, 'How do I want these things to benefit me?' You're deciding, 'Oh God, how do I manage this handful of land mines?' And your opponent is thinking the same thing but you don't know that."

The game is also designed so that playing particular events at the wrong moment can result in players unintentionally tipping the world into thermonuclear war. This is very much intentional, says Matthews. "The Cold War was very, very dangerous. The world came very close to destroying itself for stupid reasons. It's obviously true that we managed to get through it without a major nuclear exchange but there are at least four or five incidents where things got perilously close and the question that the game poses is: 'Was the peaceful outcome the most likely outcome or the only outcome?' I think the answer to that was no."

It is the dicey nature of the clash between the two nuclear superpowers that Gupta believes Twilight Struggle reveals about the Cold War. "I hope people get a sense from the game of how both superpowers really threw their weight around, how they thought of the rest of the world as basically sections of their backyards, and how that mentality is a dangerous one," he says.

While the game models a historic conflict, Matthews believes understanding the Cold War is crucial to understanding present-day geopolitics. "The current generation that is governing grew up in this context and I still think that the American people as a whole analogize back to the Cold War," he says. "It's a danger of history by analogy. Sometimes we decide the Chinese are going to be the new threat and it's going to be the Cold War all over again with the Chinese, or we're like, 'Well, the Russians are at it again.' I think those are dangerous analogies but they are ones that occur very commonly in public policy debate in the United States."

Twilight Struggle proved to be not just revealing but popular too. War games had been in the doldrums for years, played by an ever-shrinking group of die-hard fans, but since its release in 2005 Twilight Struggle has become a cult hit. For years it was rated as the best game of all time by users of the online board gaming mecca BoardGameGeek and sold tens of thousands of copies—an enormous amount for what is at heart a complex war game.

Twilight Struggle's success in bringing the Cold War back to life on kitchen tables reflects board games' ability to make complex events intuitively understandable, says Gupta. "Both board games and video games have enormous power to induce empathy, to induce players to step into someone else's shoes," he says. "You're not just reading about someone's thoughts and decisions, you are making the decisions. So the nature and texture of those decisions is what causes people to achieve a deep realization and understanding."

While Twilight Struggle's model helps players understand the defining struggle of the second half of the twentieth century, a game it inspired is doing the same with the conflict that has defined the early twenty-first century: the battle between secular democracy and Islamist jihad.

That game is Labyrinth: The War on Terror 2001–? and it is a game built on its author's own deep understanding of the subject. For while Volko Ruhnke makes board games he is also a CIA national security analyst. And while discussing the ins and outs of his day job is off-limits, it is clear that the theories behind the game would be familiar to most people wandering the halls of Langley.

In its standard scenario, Labyrinth begins in the immediate aftermath of the 9/11 terrorist attacks. The jihadists control Afghanistan and U.S. troops are stationed in Saudi Arabia and the Gulf States. The immediate question is how will the United States react?

Like Twilight Struggle, the action is driven by cards representing events such as the United States using drone strikes to take down terrorist cells to jihadists undermining the West's ability to influence Pakistan by assassinating Benazir Bhutto.

As in real life victory looks different to the two sides. The United States needs to foster stable governments in the Muslim world or wipe out every jihadi cell. The Islamists win by creating a caliphate or getting the United States to withdraw from the Islamic world, by either damaging its international prestige or carrying out a terrorist attack on U.S. soil using a nuclear or biological weapon.

"One of the premises of Labyrinth is that global jihadism is a movement," says Ruhnke. "There are local aspects to it but it is a movement that's seeking, generally, the same kind of thing. They clash against each other also but, at the end of the day, they all want this Sunni Wahhabist caliphate."

Labyrinth boils down to a tug of war over governance. America and its allies are trying to pull the Muslim world toward good governance as defined by the United Nations while the jihadists want to pull it toward theocratic dictatorship.

Not that governance is the only thing that matters in Labyrinth. The European countries on its sprawling board are all deemed to have good governance and, unlike the Muslim countries, that never changes. However, the Islamist players can still recruit terrorist cells in Britain, France, and Spain.

"You have in France, Spain, and the United Kingdom sizable Muslim populations, national diasporas, and issues of assimilation that we wanted to represent in the game," Ruhnke explains. "So that's an admission that the model is not so clean and simple in real life as to say, 'Well, we have a wonderfully functioning democracy and even efficient services for the people, and nice welfare for the population, and freedom to come and go and say what you like, and, therefore, we are not going to have any problems with jihadism.'"

What the importance of governance in the game illustrates, Ruhnke adds, is that the clash between secular democracy and fundamentalist Islam is a story of insurgency and counterinsurgency. "If we think about these people as terrorists it is less illuminating in terms of what is going on than if we think of them as insurgents," he says. "If we look at it globally, as Labyrinth does, it's a global insurgency and so we are, through our coalition and so forth, trying to counter that."

Insurgency and counterinsurgency are the common themes to all the games Ruhnke makes. Since Labyrinth he has created games about conflicts as diverse as the Colombian Civil War (Andean Abyss), the U.S. campaign in Afghanistan (A Distant Plain) and the Roman Empire's battles with the Gauls (Falling Sky: The Gallic Revolt Against Caesar), but the nature of insurgency lies at the heart of all them.

Insurgency, he says, defines almost every present-day conflict from Syria to Ukraine. And, as Ruhnke's games make plain, what really matters to insurgents is velocity.

Can the United States wipe out Islamist influence faster than the Islamists can expand it? Playing as the United States in Labyrinth often feels like an elaborate game of Whac-A-Mole, where each victory is countered by another jihadist cell spawning elsewhere on the board. You might push the Islamists out of Afghanistan, but if you're bogged down there for too long, you may lack the resources to deal with them regrouping in Sudan. "One thing that is very important for insurgency, counterinsurgency is that it is very attritional guerrilla warfare. It's not these sweeping tank panzers and things like that," says Ruhnke. "What's really important is rates and flows and inputs and outputs. There's going to be casualties in attrition. The question is not so much, 'Are you killing my bases?' as 'Am I rebuilding my bases faster or slower than you kill my bases?'"

Inevitably Labyrinth, much like the military war games that helped plan the attack on Pearl Harbor, lends itself to testing what-ifs. When I played Labyrinth for the first time as the United States, I didn't invade Afghanistan and instead focused on winning the war of ideas and improving governance elsewhere. It proved to be a disastrous decision.

My Islamist opponent entrenched themselves in Afghanistan, building up cell after cell before pushing north into the former Soviet states of central Asia, where they obtained a stash of highly enriched uranium. Soon after the Islamists won the game when one of their Western European sleeper cells made it to the United States and detonated a nuclear weapon.

Yet delivering regime change in Afghanistan is a slow and draining process and the outcome depends as much on how the Islamist player reacts as what the U.S. player does.

Unsurprisingly for a game about an ongoing and politically charged conflict, Labyrinth has sparked plenty of debate about the validity of its model. "There's a review on BoardGameGeek that says it is a fun game but Volko gets terrorism wrong, which was interesting because I think I know something about the subject," he says. "It stimulated a fantastic exchange of ideas and interpretations on all sides. That's the best kind of reaction. It's not simply: 'Get your facts right.' It's about trying to refine our views and I think a game model can represent it better than a newspaper article can."

And this freedom to experiment and play with the model is what gives board games like Labyrinth, Twilight Struggle, and Pandemic the edge over other media when it comes to understanding byzantine global affairs like the war on terror. "You can explain that system to somebody but to internalize it, to get it intuitively, remember it, and carry it with you, it's much easier to have that happen if you experience that," says Ruhnke. "What's happening on the board affects you on some level that wouldn't affect you if you just had a professor explaining it to you at West Point."

Video games might seem to offer the same ability but there's a crucial difference. The rules of board games are transparent, on display to all players, rather than hidden within reams of inaccessible computer code.

"With a video game you can provide a look into the program and you can provide documentation but inevitably there will be a black box effect," he says. "What besides a manual game where players operate the model gives you that transparency? Nothing. Nothing!"

16

# MADE IN GERMANY: CATAN AND THE CREATION OF MODERN BOARD GAMES

## *How Germany revitalized board gaming for the twenty-first century*

Outside the convention center thousands of people are lining up. Inside, the crowds have made a hall big enough to hold three football fields feel like a rush-hour commuter train.

At exhibition stands, people are handing over wads of euros for board games and sitting at tables playing the latest releases. The only stand that doesn't have people buzzing around it is the one selling video games. Everywhere there are people dragging carts loaded with board games behind them, debating where to go next, and chowing down on currywurst and bratwurst.

At the Hasbro stand a young girl is squealing with glee after watching her father get a face full of confetti while playing Pie Face, a plastic action game that dares players to see if they can turn the switch without getting splattered. In the next hall players are asking Matt Leacock to sign copies of his game Pandemic. In yet another hall three Frenchmen are learning how to play €uro Crisis, a new board game where players asset strip debt-ridden European nations.

As the players discuss selling off the Greek islands, a couple dressed as comic book antihero Deadpool and *Star Wars*' Princess Leia wander past. A good portion of the crowd are men wearing black T-shirts promoting death metal bands like Bolt Thrower and Carcass but they, and the dressers-up, are in the minority. Most of the people here wouldn't stand out in a shopping mall. There are young couples holding hands, elderly women with walking canes, and families enjoying a day out with the kids.

Away from the halls, in the bars and hotels surrounding the convention center, publishing company representatives are testing prototypes and striking deals for games that the crowds will play in years to come.

Welcome to Essen, Germany, home to Spiel—the world's biggest board game convention.

Spiel began in 1983 as a weekend of game playing in a local school and from that humble beginning it has grown into a four-day event that attracts more than 150,000 people every year. Each October Spiel fills most of the enormous Messe Essen center and makes hotel rooms in the city a scarce commodity.

In 2015 Spiel's nine hundred exhibitors swallowed up 75,000 square yards of floor space, and around 850 new games were on display, many of which were making their public debut.

If any one factor explains how Spiel became the spiritual home of board gaming, it can be found in the hangar-size Hall 4. For most of Spiel 2015, the entrances to Hall 4 have been sealed, but at 4:30 p.m. on the Saturday the shutters rise and more than one thousand people pour in.

Inside are row after row of pushed-together tables that stretch from one end of the hall to the other. In front of the folding chairs lining the tables is a paper game board depicting an island made out of hexagons, fresh card decks, and transparent bags filled with plastic playing tokens.

The game on the table is Catan, a contest of resource trading and settlement building that's become a phenomenon. Since its German debut in 1995 as Die Siedler von Catan, the game and its four expansions have sold more than twenty-two million copies worldwide and rejuvenated tabletop play.

All of the 1,040 people in Hall 4 are here to play one giant game of Catan together. Each board is connected to those around it by the seas surrounding each island and the first player in the hall to earn twenty-five points from developing their settlements will win the game.

Over the next hour and a half the hall becomes a frenzy of deal making against the clock that resembles the days of stock market pit trading. "Wool! I have wool! Wool for brick!" shouts one player. "Need lumber! Anyone have lumber to trade?" asks another, who wants the wood to build a new settlement. "I will swap one lumber for two ore," offers the man next to her.

In between the bouts of wheeling and dealing, the game's host announces the moves for the robber, the bandit that roams each island and prevents people earning new resources from the spaces he lands on. Every time the robber pounces on a new resource half the hall groans in agony as their precious supplies are lost while the other half breathes a sigh of relief that the thief is no longer stealing from them.

Eventually a Dutch player signals that he has enough points to claim victory. The game freezes instantly. A referee scampers over to the player's board and starts double-checking the score as the hushed hall waits for the verdict. The referee looks up and signals that the score is correct. "We have a winner," declares the host.

As the player heads to the stage to collect his prize from Catan creator Klaus Teuber, the hall applauds. There is no animosity or envy here. This isn't a competition, it's a celebration—a celebration of more than one thousand strangers, many of whom do not even share a language, having fun together and jointly smashing the record for the biggest-ever game of Catan.

Catan grew out of Teuber's lifelong obsession with Vikings. As a child he owned a set of beautifully painted plastic Viking figures and the more he read about medieval Scandinavia the deeper his fascination with the Old Norse grew.

Far from being the raping and pillaging raiders of popular imagination, Teuber discovered that the Vikings were great craftsmen, bold adventurers, and imaginative storytellers. Their trade links extended so far that they were even trading with Baghdad, which they reached via the Volga, the great river that brought chess from Persia to Russia.

He discovered that the Vikings were the greatest seafarers of their time. While others hugged the shoreline, the Vikings sailed into the open ocean and founded settlements on Iceland, Greenland, and Newfoundland. They were also democrats who made their laws in community assemblies where all free men had a say.

As for those famous horned helmets, they were a myth, invented by nineteenth-century Europeans who wanted to highlight Viking marauding, even though their taste for raiding was hardly exceptional by the standards of early medieval Europe.

"He is a huge fan of Viking history," says Klaus's son Benjamin Teuber, who now works with his father on Catan and other games. "He always dreamed of exploring new lands. As a kid he always dreamed about it and his teachers said that he will never achieve anything in his life if he continues to dream like this."

Klaus never stopped dreaming. While his family did not play board games he became drawn to them in his early teens and started making his own. "Then I lost my interest because of girls and other more important things," he says with a smile. But after becoming parents in the early 1980s Klaus and his wife started playing games together. "We felt each evening of television was not so good and was boring so we played games," he says.

Interest rekindled, Klaus began making games again. The motivation was fun, not money. "I didn't know there was a possibility to do that," he says. "I thought companies made them and I didn't think there were authors."

One day he showed some people at a German games convention a game he had designed. It was called Barbarossa and involved players having to deduce the meaning of one another's modeling clay sculptures by asking yes or no questions. The people he showed it to put him in touch with Kosmos, the Stuttgart game publisher that went on to release Barbarossa in 1988.

More games followed. Some won awards and sold well in Germany but none made enough money to convince Klaus to quit his job as a dental technician. Then in the early 1990s changes to German health-care policies pushed his dental practice to the edge of bankruptcy. As the stress at work mounted, Klaus immersed himself in his games. "In the evening it was my refuge, my escape, my own world, where no one could speak to me," he says. "It was something that was my world."

Buried in his game making, Klaus turned once more to the adventures of the Vikings. "They discovered new lands that had no inhabitants like Iceland," he says. "That sparked my imagination. What would they do there? So I started to build this picture in my brain about that discovery as a game."

Klaus's original vision for Catan was much broader than the game eventually released. As well as building and trading, he imagined players sailing the seas to find new lands, fleeing pirate ships, and even fusing settlements together to create medieval metropolises. With so much going on, the early versions were a drag to play. "I think 1992 was the first time I played it, or 1993, I was like eight or nine," says Benjamin. "It wasn't good at all; it was very complicated."

But in the years that followed Klaus edited the game down. He chopped out feature after feature, whittled down the range of resources, and repeatedly polished what remained until he had distilled Catan into a streamlined experience that was neither too simple nor too complex. Its roll-trade-buy action took mere minutes to learn but the ebb and flow of its tabletop economy and the ability to reassemble the hexagonal land tiles to create new islands meant Catan never felt stale.

Good as it was, few suspected in 1995 that Catan would start a revolution. Not even the game's publisher Kosmos. "I knew the publishing house and the guy who was responsible for it at the time didn't expect such a success," says Dominique Metzler, head of Spiel organizers Friedhelm Merz Verlag. "They thought: 'It's a strategic game, we won't reach that many people with it.' They didn't really think about reaching families and stuff like that."

Even after Catan won the prestigious Spiel des Jahres, Germany's annual game of the year award, everyone in the business thought it would sell well for a year and then be out of print within three to four years. After all, Klaus had won that award three times before and that's exactly what had happened.

One reason few in the German games business thought Catan would be extraordinary was because it wasn't. Not for Germany at least, for—largely unnoticed by the rest of the world—the country had been forging its own path in gaming and elevating board games from mere distractions to something closer to precision-engineered art.

Elsewhere board games had stagnated. Enduring favorites like Clue and Parcheesi had become the equivalent of aging sedans, popular and widely used but hardly thrilling experiences. The main alternative to these family hits, however, were complex strategy games about war and business—the oversized, over-engineered Hummers of gaming that came with a blizzard of plastic figurines and rulebooks thick enough to use as doorstops.

But by the time Catan arrived, German games had become sleek BMWs. They offered finely tuned experiences that put decision making and strategy ahead of luck but did so with elegant rules and short play times. German games also rejected the dog-eat-dog competition common in U.S. games in favor of experiences where player cooperation was crucial.

The foundations of the Germans' tabletop exceptionalism were laid in the aftermath of World War Two as West Germany rebuilt not just its war-ruined cities but also its culture. In the postwar period the Germans made board gaming part of everyday life.

"Before the war children played and adults didn't play games," says Tristan Schwennsen, archivist for Germany's biggest game publisher Ravensburger. "Games were for children and it was very important that children learned while they played the games, learned something about the world or other countries. After World War Two it changed because people got more leisure time."

As West Germany revitalized itself, board games became associated with togetherness and wholesomeness; a feeling reinforced by newspapers that reviewed board games alongside the latest novels, movies, and stage plays. The sense that playing games was good and social eroded away the belief that tabletop gaming was just for kids and turned West Germans into the world's biggest consumers of board games per person.

Tellingly, this change in attitude didn't take place in communist East Germany where leisure time grew more slowly than in the free-market west. Klaus recalls that in the first game fairs held in the east after reunification, mothers would arrive expecting to be able to leave their kids at the show while they went shopping because they assumed it was only for children.

Today, however, board gaming is part of life throughout Germany. "Everyone plays," says Schwennsen. "In Germany we have games for everyone. Games for geeks, games for children, games for older people, games for families. Everyone plays. No one makes fun of anyone who plays board games, they are not thought of as childish or anything like that."

Benjamin agrees. "Being in that culture you consider it being normal. I have a good friend from Spain and I told him I work with board games and he said: 'That's just for kids.' That's their perception of it," he says. "But in Germany it really is like: 'What will we do tonight? We don't really want to go out tonight, it's cold outside, it's rainy, but we want to do something other than just watching TV.' And then you say: 'Well, why don't we play this board game?' That's something which is in the culture."

As well as embracing play by all ages, postwar Germany turned decisively against the use of war in entertainment. Violent toys and games became taboo and anything that seemed to promote militarism faced censorship. When Parker Brothers tried releasing Risk in West Germany, the government's censors threatened to restrict its sale, forcing a rewrite of the rules so that players were liberating rather than conquering the world.

"I think, based on the horrible German history that we all have, somehow now the people have developed a sense that we need peace," says Benjamin. "Now the mind-set is very liberal and very freedom-orientated because we know that something like seventy years ago can never happen again. Very few war games come from Germany, they mostly come from the States or other countries. Most German games are really cooperative."

For anyone raised on a diet of Monopoly and chess, the lack of direct competition in Catan and other German games is striking. Catan players can only achieve victory by trading with others. There are no armed struggles and players are never forced, through poverty, to hand over their property. In Catan everyone depends on one another. Even the winners stand on the shoulders of others.

"I did not want any war in the game because it's something we have enough of in the world and I think you should not experience that again in a game," says Klaus. "Some people wrote me emails saying: 'We want to have the ability to destroy the villages or cities.' But no, not in Catan."

But changing attitudes to games and a desire to avoid conflict were no guarantee that Germany would develop its own style of board game. The real turning point came in February 1978 when a bunch of journalists got together after a day spent reporting on the Nuremberg toy fair.

Every year the journalists met to eat, drink, and talk shop at the apartment of Tom Werneck, a writer who lived in the city of Erlangen, just north of Nuremberg. They spent much of the evening discussing the lack of media attention given to new games and how this meant people were missing out on the very best releases.

Eventually Jürgen Herz, who worked for the tax-funded broadcaster WDR, made a proposal: Why don't we play every game released this year and then choose the game of the year, the Spiel des Jahres, and use that award to spread the word about the best game?

By the end of the night the journalists had agreed to give Herz's idea a try. Over the following months they laid out the rules. To make the prize credible, the jury would need to be made up of experienced games journalists who must never do any work for the games industry. To make the award relevant, only games published in German would be eligible and while the award would recognize innovation it would also seek to recommend games that anyone could enjoy rather than just hobbyist gamers.

Rules in place, the jurists spent a weekend playing every game released in 1978 and after much arguing settled on a winner: Hase und Igel, a game first published in Britain in 1974 as Hare and Tortoise that had just been released in Germany by Ravensburger.

Although the German edition was named after the Brothers Grimm's fairytale *The Hare and the Hedgehog,* the game's English designer David Parlett had based it on Aesop's fable about the race between the tortoise and the hare. Hase und Igel reinvented the race game. Instead of movement being governed by the rolling of dice, players moved by spending carrot cards that acted like fuel. Since moving fast required spending more carrots per space and players could not cross the finish line with more than twenty carrots in their hand, the game was all about math and planning ahead rather than luck.

Hase und Igel's reliance on spending carrots wisely meant that even players who seemed to be lagging behind in the race could still make a comeback and snatch victory, making for a less predictable and more strategic experience. This focus on planning over chance provided players with an early taste of how German games would develop over the next decade.

Despite picking their winner the journalists never got round to telling anyone that Hase und Igel was the 1978 Spiel des Jahres. "These guys were not very well organized in the beginning and so they elected a game and they were unable to organize an event to show the media what game they had elected," says Tom Felber, the current chairman of Spiel des Jahres. "So they elected Hase und Igel again in 1979 and then they were able to invite the public to show what they elected."

After overcoming its false start, the Spiel des Jahres award steadily gained more and more influence. Year after year it delivered its verdict and each time its logo—a red pawn wearing a golden laurel wreath—became a bit more recognized by the German public and a bit more coveted by game publishers and authors. Over time Spiel des Jahres winners started to become instant bestsellers in Germany and publishers began paying for the right to stick its logo on the packaging of winning games.

Today the Spiel des Jahres is a mark of quality recognized by most Germans. Winning it can increase sales of a game in Germany by hundreds of thousands of copies.

"In Germany it is really powerful because people trust it," says Felber. "People don't have to think anymore at Christmastime, they can just go into the shop and buy the game blindly and they always know it is a good choice."

The award became a catalyst for the emergence of games that were innovative but still had mass appeal. German game designers and publishers began upping the quality of their games in the hope of winning the accolade and the huge sales that came with it. As word of the prize's sales boost spread, game designers in other parts of Europe followed suit, hoping to score a hit in board game–hungry Germany.

By the mid-eighties German game designers were breaking away from the roll-and-move mechanic that had dominated board games for centuries. Just as Kraftwerk had dismantled pop music and reassembled it into a new exciting form with their 1975 hit "Autobahn," the Germans were engineering the games of the future.

German games fizzed with fresh thinking. In 1983's Scotland Yard one person became the unseen fugitive on the run in central London while the other players became the detectives trying to follow the trail of clues and trap the criminal. The following year's spy-themed Heimlich & Co. added bluffing to the roll-and-move mechanic so that players had to try and conceal their own piece's identity from others by moving multiple tokens on their turn. Manhattan, the 1994 Spiel des Jahres, used the concept of auctions to create a game where players vied to dominate cities by adding floors to spindly 3D skyscrapers that rose from out of the game board.

While the Germans were enjoying this burst of Teutonic creativity, Americans faced store shelves dominated by decades-old family favorites, Trivial Pursuit clones, and tepid efforts like Hotels, which amounted to little more than Monopoly with random planning permission and fancy cardboard buildings.

By the mid-nineties a common set of principles could be seen in the new wave of German games. The central principle was that the outcomes of a game should be governed by player decisions rather than fate. Dice were used sparingly and sometimes not at all. Cards often added a degree of randomness, but the passive "spin and see who wins" approach of the Game of Life was now deemed dull and dated.

Instead of chance, players would be given a small range of options for what to do on their turn and their success in the game rested on how good their decision making and planning were relative to other players'.

"Germans like to make plans and the typical German board game is all about strategy," says Heinrich Hüentelmann, head of media relations for Ravensburger. "If it's all about strategy you are better off when you know how to make your plans, and when you look at complicated and complex things it is better to make a plan and look at the future."

The Germans also wanted streamlined rules, preferring games that offered more complexity than Monopoly but stopped well short of the time-consuming detail of American strategy games, such as the World War Two–themed Axis & Allies with its abundance of plastic miniatures, rules, and combat tables.

And since games were seen as a social activity, German games involved everyone right up until the end. Games like Risk or Mouse Trap, where players could be eliminated before the game ended, became frowned upon.

"In American-style board games players are eliminated, like in

Hotels—so you have no money, go, we will carry on playing, you go do the dishes or whatever," says Schwennsen. "Never ever would you find such elimination in a German board game because it is important that all people are playing together. Of course you want to win but it is important that you are all at a table and spend time together. To eliminate a player doesn't fit in with that idea."

This desire to keep everybody playing has also led to a belief that games shouldn't allow front-runners to emerge. The all-too-common Monopoly situation where one player is clearly going to win but the game drags on for another half an hour until everyone else is bankrupt was out. Instead German games allowed last-minute comebacks or made it impossible to be sure who was ahead by concealing each player's goals or their progress.

These principles were not exclusive to Germany by any means. Hase und Igel embraced many of the same ideas despite being made in Britain, as did Acquire—a refined 1962 game of hotel mergers and acquisitions created by the U.S. game designer Sid Sackson. But these were isolated efforts. With its "everyone plays" culture and the Spiel des Jahres, Germany had the commercial and quality incentives necessary to cultivate a game design movement built around the shared ideals of streamlined rules, decisions before luck, no player elimination, and vagueness about who is winning.

Catan embodied all of the emerging design principles of what would later be clumsily labeled as "German-style games" or "Eurogames" by Anglosphere players. In Catan players did not fight directly, no one got eliminated, and planning and trading trumped luck. The rules were easy to learn, the race for victory was tight enough for anyone to feel they could win throughout, and the average game was over after an hour. Catan hadn't pioneered these concepts, but few other games had delivered on the new ideals of German game design with such flair.

Catan won the 1995 Spiel des Jahres and sales inevitably soared, but then something strange happened. Instead of the game's sales slipping in 1996, they rose again. The next year sales rose yet again and

Klaus finally found himself in a position to quit dentistry and make games full-time.

Alan R. Moon, an American game designer who was working for Ravensburger in the late 1990s, remembers Catan's popularity was unlike anything seen before in Germany. "I was in Ravensburg one day and, with several other executives, went to a supermarket at lunch," he recalls. "At the checkout counter there was a steel bin of Catan. I looked at the guys from Ravensburger and asked: 'Have you guys ever had a game in the supermarket?' They all went: 'No, we haven't.' So that tells you how big that was, it was just huge."

Even bigger success was to come.

By the time of Catan's release, few of the games making waves in Germany had ever made it to North America. So rarely did these games get released outside Europe that even those working in the U.S. games business struggled to get hold of them. Moon first got interested in German-style games in the early eighties while working for strategy game specialists Avalon Hill, but he had to resort to swapping games by air mail with contacts in Britain to get copies.

These transatlantic exchanges eventually led to Avalon Hill releasing a few German-style games in the United States including Heimlich & Co., which was published under the name Under Cover, and Kremlin, a parody of Soviet politburo backstabbing first released in Switzerland.

But these releases were few and far between. Photocopied fanzines with few readers became the only source of information about German games in the United States, and readers tempted to buy these games faced paying for expensive imports with instructions written in German.

One supplier of these imports was Chicago's Mayfair Games, which began bringing German games to the States in the early nineties, mainly because founder Darwin Bromley loved them. After several years of this, an employee named Jay Tummelson persuaded Bromley to start producing English-language editions of the best European titles, and in 1995 Mayfair began buying the rights to German games. One of the first games the company signed was Catan.

Around the same time the Internet was bringing together the board game fans who had previously relied on fanzines for information. On the message boards and online newsgroups, word spread about the great but hard-to-find games being released in Germany, and one game in particular was raved about more than any other: Catan.

So when Mayfair Games released Catan in America there was already a small audience desperate to get their hands on it. The online buzz about German-style games grew even more when Tummelson left Mayfair Games and started Rio Grande Games, a publisher dedicated to bringing Eurogames to North America.

By the early 2000s Eurogames were all the rage in U.S. board game clubs and new publishers such as Z-Man Games and Days of Wonder began forming to tap into the new interest. "We played a lot of new Eurogames," recalls Twilight Struggle cocreator Ananda Gupta, who belonged to a George Washington University board game club around this time. "Games would rarely be played more than once or twice; we were always chasing the new. The group seemed to have a fascination with games that had camels in them so a lot of games involving camels were played. We had all played ourselves out of Catan by then—I played something like four hundred games of Catan at college. At this time everything was still called by its German name because the games were not as ubiquitous."

For the Germans the growing excitement among English-speaking hobbyists was somewhat mystifying. "Someone found these strategic games, this word 'German game,' and all of a sudden everybody from abroad who was interested in games wanted to play these kinds of games and that led to this big term 'German board games,'" says Spiel chief Metzler.

Schwennsen agrees: "To compare the games we have after Catan and before Catan, I wouldn't say there is much difference. Just because German board games are interesting for people all over the world doesn't make a difference for us because we had these games before."

These games might have been the norm for Germany, but for board game aficionados in the United States each new discovery was treasured like a gold nugget.

Soon the Eurogames began influencing North American game designers. "I grew up in the United States, so as a kid until the eighties, game invention for me was very U.S.-centric," says Mike Gray, a former Milton Bradley game designer whose credits include the 1986 war game Fortress America and the 1989 shopping spree game Electronic Mall Madness. "But from the late eighties European games started to be more popular, they started to sort of be brought over here and translated to English, and that slowly changed the way people thought about games."

Alan Moon was among the designers who embraced the European approach, and in 1998 he won the Spiel des Jahres with Elfenland, a game where players helped elves travel from city to city on unicorns, magic clouds, dragons, and other fantastical modes of transport. He is one of just three non-German game designers to have won the prize since 1986.

U.S. board game designers also began traveling to Essen to join the crowds at the Spiel convention. Among them was Matt Leacock, who made his first trip there several years before creating Pandemic. "Spiel was like nothing else," he recalls. "I had been to [the U.S. game convention] GenCon before but that felt much more role-playing centric. It was exciting to go to GenCon but I also felt like it was an adjacent group of people—their hearts were in role playing. When I went to Spiel it was amazing because it was all the types of games that I loved and it was everywhere—it was transformative. The fact that there were so many people there too was just glorious."

Despite all the excitement among American game designers and board game fans, Eurogames were no more than a cult pursuit. Even by 2004 Catan had yet to sell more than one hundred thousand copies in the United States, a feat the game managed several times over in its first year on sale in Germany almost a decade earlier.

But as the years went by sales of Catan kept accelerating. By 2008 more than half a million copies had been sold in the United States and then in 2009 the game finally broke into the mainstream after the movers and shakers of Silicon Valley embraced it.

Catan became the game to be seen playing among the tech set during 2009. LinkedIn founder Reid Hoffman and Mozilla chief executive John Lilly raved about Catan. Facebook started hosting Catan competitions for staff working at its Palo Alto headquarters. Mark Pincus, the chief executive of FarmVille creators Zynga, told *The Wall Street Journal* that Catan was Silicon Valley's equivalent of executive golf: "None of us have time to play eighteen holes of golf, but we can handle a pizza and a board game."

The timing could not have been better, coming right at the peak of global excitement about social networking and the iPhone. Silicon Valley was cool and had yet to become associated with tax avoidance, invasions of consumer privacy, and political lobbying. So when the news broke that the big hitters of big tech loved Catan, sales soared.

"It was the best marketing for us," says Benjamin. "Afterwards came the actors, like Reese Witherspoon and Mila Kunis, then came the sportspeople—quarterback Andrew Luck said it is his favorite game—and now even models say they love to play Catan. It started from this geek chic area, the techies in Silicon Valley, who weaved it into the cultural fabric and into a more mainstream thing."

Soon after Catan appeared in TV shows from *Parks and Recreation* and *Big Bang Theory* to *The Simpsons* and *South Park*.

By then Catan wasn't the only German-style game making an impact in the English-speaking world. The infection-fighting cooperative game Pandemic was growing fast, as was Carcassonne, a German game released in 2001 where players create medieval landscapes from tiles while trying to score points by claiming cities, roads, and fields.

Another emerging hit was Moon's second Spiel des Jahres winner Ticket to Ride, a 2004 game where players jostled for control over the railroads crisscrossing North America. Moon got the idea for Ticket to Ride one crisp spring morning when walking along the Atlantic coast in Beverley, Massachusetts. "It was just one of those 'aha!' moments," he says. "When a game designer works out a game in their head they all work wonderfully and you think it's going to be the next big thing, but when you get the prototype out and actually play it a huge percentage of them aren't very fun. But this was the exception. We played it the first time and I was like, 'Wow, it works great.'"

His confidence only increased when he invited friends over to play the prototype. "I was actually playing the prototype early on with a few friends and I got up from the table and a splinter went into my leg from the table," he says. "I went into the bedroom to take my pants off and see how big it was and this four-inch-long splinter had gone completely through the top of my leg. So I came back out and said: 'You know what I've got to go to the hospital, I've got to get this splinter out, you guys just keep playing.' When I came back two hours later they were still playing the prototype and having a good time."

After selling the game to Days of Wonder and winning the Spiel des Jahres, Moon expected Ticket to Ride to have a good year before sales dried up. Instead, just like Catan, it kept selling. Then Days of Wonder created smartphone versions of the game and, to Moon's horror, decided to give it away for free every so often.

"Days of Wonder's philosophy was that they didn't mind giving it away versus charging a dollar ninety-nine or whatever because they figured a certain percentage of those players would go out and buy the board game, which is where they make the big profit," he says. "I thought that's terrible, you're giving the game away for free, but they were absolutely right."

Ticket to Ride went on to sell more than three million copies worldwide.

Together these German-style games became the first blockbuster board games of the twenty-first century, selling in quantities comparable to established classics like Clue and Risk. "Being a game designer is like being an artist in terms of your chances of success," says Moon. "In the last twenty years there's been Catan, Carcassonne, and Ticket to

Ride; those are really the only mass-market games in the U.S. of that group. That's three games in twenty years. That's what your odds are."

Although most of the new board games never go mainstream, they have—collectively—rejuvenated board gaming. In 2014 sales of hobby games, as the industry likes to pigeonhole them, were up 20 percent over the year before, bringing their annual U.S. sales to $880 million.

The passion these games have inspired among players is equally impressive. Between the summers of 2009 and 2015 users of the crowd-funding website Kickstarter pledged $196 million toward the development of new board and card games. Video game projects lagged behind with pledges worth $179 million.

Today the design principles honed in Germany during the eighties and nineties are even influencing the development of board gaming perennials, with Monopoly Empire abandoning player elimination and Risk: Star Wars Edition requiring players to plan ahead and balance competing priorities.

Boldest of all was Rob Daviau's Risk Legacy. Released in 2011, Risk Legacy transformed the popular war game into an on-going campaign spread over multiple games. During the game players would be required to rip cards in two and apply stickers to the board that would permanently change how the game worked. After fifteen games the campaign ends and the board stops changing but what is left is a unique game that doubles as a diary of the players' past games.

"Games are coming from anywhere and everywhere now, which is really exciting," says Jonathan Berkowitz, Hasbro Gaming's senior vice-president of marketing. "There's a much more robust gaming community, a lot more new games are created now than ever before."

Berkowitz thinks that part of the reason for the current board gaming boom is due to the growing numbers of adult gamers. Three decades ago board games were viewed as "for children" but today adult players abound. This appears to be part of a wider, generational shift in society's attitude to play that may well have begun with the excitement over Trivial Pursuit in the mid-eighties but was probably also encouraged by the success of video games.

Back in the eighties, video games, just like board games, were seen as the preserve of Mario-loving kiddies. But since the nineties, the video game industry has been delivering more and more games for adults that have kept millions from putting down their game pads on reaching maturity. Many of the biggest video games of today, from Grand Theft Auto and Resident Evil to Borderlands and Call of Duty, are now designed with adult players in mind, even if they do still find their way into the hands of minors.

The Eurogames have done much the same for board games, broadening tabletop gaming's appeal to those who want to keep playing after childhood with new experiences and themes that connect with adults better than family favorites like Mouse Trap, Candy Land, and Sorry!

Indeed, it is hard to imagine that Twilight Struggle's Cold War strategizing would have found an audience outside the war gaming niche if it wasn't for Eurogames encouraging adults to try new types of game. Nor would it have seemed likely that Agricola, a tense game about raising a family of seventeen-century European farmers and avoiding starvation, or Power Grid, where players compete to supply electricity to German and U.S. cities, would have become cult hits.

And board games based on fresh thinking keep coming too—from *The Walking Dead*–echoing Dead of Winter, where players are a group of people with hidden agendas trying to survive a zombie apocalypse, to the Spiel des Jahres–winning Camel Up, where players bet on unpredictable dice-governed camel races.

Not that the resurrection of board gaming is just about Eurogames winning over adult players and inspiring the opening of all those board game cafés. There are deeper cultural trends at work too, says Berkowitz, who notes that sales of traditional board games like Monopoly and Operation are also on the upswing.

"It's something of a counter-trend to digital and digital gaming," he says. "Our data is definitely showing that people want to get together and parents especially want to connect with their kids and kids want to connect with their parents. It's just such a busy world out there that they sometimes have a hard time making the time and games help make that time. I don't think it's people feeling they are losing something so much as we need to balance out the equation and everyone knows that family and friends are the most important thing that we have."

And it is this ability to bring us together face-to-face that is tabletop gaming's secret weapon and what ensures that, even in an age of smartphones and PlayStations, board games are thriving rather than dying.

Given that they have been with us for millennia, board games should feel stale and outmoded by now, but the reality is they haven't felt fresher or been so popular for decades. And why shouldn't they? Board games are not and have never been mere distractions.

From the spiritual guidance offered to ancient Egyptians by senet and the stains of history on the chessboard, to the misunderstood message of Monopoly, and Catan's message that none of us can succeed alone, board games continue to evolve to reflect our needs and desires and our outlook on life. For now we may favor the games that suggest we are the agents of our own destiny and shun the fatalism of luck-based games, but in fifty years' time we may disagree.

Whatever we choose and wherever the future takes us, board games will be there, bringing us together and mirroring our choices and our attitudes on paper and cardboard.

# REFERENCES

### Introduction: The Birth of a New Gaming Era


AFP. "Board Games Cafes Offer Web Break in China." *The Independent,* February 20, 2011, www.independent.co.uk/life-style/board-games-cafes-offer-web-break-in-china-2220369.html.

Cathryn. "Interview With 'Fish Men' Playwright and New Yorker Cándido Tirado." *Washington Square Park Blog* (blog), May 4, 2012, www.washingtonsquareparkblog.com/2012/05/04/interview-with-fish-men-playwright-and-new-yorker-candido-tirado-play-features-wsp-chess-area-a-recollection-of-chess-plaza-hey-day-healing-after-911-more.

Fraiman, Michael. "Toronto Board Game Café Snakes and Lattes Gets Its Own Sitcom." *Globe and Mail,* July 5, 2015, www.theglobeandmail.com/news/toronto/toronto-board-game-cafe-snakes-and-lattes-gets-its-own-sitcom/article25275558.

Hines, Lael. "Chess Moves: Most Players Are Now at Union Square." *The Villager* (New York), August 8, 2013, thevillager.com/2013/08/08/chess-moves-most-players-are-now-at-union-square.

Lavender, Dave. "Tabletop Games Enjoying Resurgence." *Herald-Dispatch* (Huntington, WV), January 9, 2016, www.herald-dispatch.com/features_entertainment/tabletop-games-enjoying-resurgence/article_cb89774f-0876-5d80-9144-a41b6fa2665a.html.

McClain, Dylan Loeb. "In Street Chess Games, a Pedigreed Pastime Becomes a Gritty Sideline." *New York Times,* September 17, 2007, www.nytimes.com/2007/09/17/nyregion/17hustlers.html?_r=3&oref=slogin.

NPD Group, The. "Annual Sales Data." Toy Industry Association, May 2016, www.toyassociation.org/tia/industry_facts/salesdata/industryfacts/sales_data/sales_data.aspx?hkey=6381a73a-ce46-4caf-8bc1-72b99567df1e#.WCWcljKcZBw.

Ohrstrom, Lysandra. "Local: Rival Thompson Street Chess Clubs Remain in Middle Game." *Observer* (New York), April 4, 2008, observer.com/2008/04/the-local-rival-thompson-street-chess-clubs-remain-in-middle-game.

Sathe, Gopal. "Lounge Review: The Mind Cafe, Delhi." *Live Mint,* January 6, 2012, www.livemint.com/Leisure/dJC6Bef73mYRfUC2mwiLQO/Lounge-Review-The-Mind-Cafe-Delhi.html.

Schank, Hana. "How Board Games Conquered Cafes." *Atlantic,* November 23, 2014, www.theatlantic.com/entertainment/archive/2014/11/board-game-bars/382828.

Vasel, Tom. "Games in Korea." *Games Journal,* January 2004, www.thegamesjournal.com/articles/GamesInKorea.shtml.

### 1. Tomb Raiders and the Lost Games of the Ancients

Beitman, Bernard D. "Brains Seek Patterns in Coincidences." *Psychiatric Annals* 39, no. 5 (May 2009).

Bell, R. C. *Board and Table Games from Many Civilizations.* New York: Dover Publications, 1979.

Carter, Howard. *The Tomb of Tutankhamun Vol. 3: Treasure & Annex.* London: Bloomsbury Academic, 2014.

———. "Tutankhamun: Anatomy of an Excavation." Griffith Institute, October 6, 2010, www.griffith.ox.ac.uk/gri/4sea1not.html.

de Voogt, Alex. "Distribution of Mancala Board Games: A Methodological Inquiry." *Board Game Studies* 2 (1999).

———. "Makonn and the Indian Ocean: East African Slave Trade and the Dispersal of Rules." *Board Game Studies* 8 (2014).

de Voogt, Alex, Anne-Elizabeth Dunn-Vaturi, and Jelmer W. Eerkens. "Cultural Transmission in the Ancient Near East: Twenty Squares and Fifty-Eight Holes." *Journal of Archaeological Science* 40 (2013).

Dodson, Aidan. "Egypt: The End of a Civilization." *BBC History,* February 17, 2011, www.bbc.co.uk/history/ancient/egyptians/egypt_end_01.shtml.

Earl of Carnarvon and Howard Carter. *Five Years' Explorations at Thebes: A Record of Work Done 1907–1911.* Oxford: Oxford University Press, 1912.

Faulkner, Raymond, et al. *The Egyptian Book of the Dead: The Book of Going Forth by Day: The Complete Papyrus of Ani.* San Francisco: Chronicle Books, 2008.

Finkel, Irving. "On the Rules for the Royal Game of Ur." In *Ancient Board Games in Perspective,* edited by Irving Finkel. London: British Museum Press, 2007.

"Games New Yorkers Play: An Exhibition of Games." *Horizon,* January/February 1985.

Gobet, Fernand, Alex de Voogt, and Jean Retschitzki. *Moves in Mind: The Psychology of Board Games*. Hove, UK: Psychology Press, 2004.

Green, William. "Big Game Hunter." *Time,* June 19, 2008, content.time.com/time/specials/2007/article/0,28804,1815747_1815707_1815665,00.html.

Kronenburg, Tom, Jeroen Donkers, and Alex J. de Voogt. "Never-Ending Moves in Bao." *ICGA Journal* 29, no. 2 (June 2006).

Maitland, Margaret. "It's Not Just a Game, It's a Religion." *The Eloquent Peasant* (blog), October 14, 2010, www.eloquentpeasant.com/2010/10/14/its-not-just-a-game-its-a-religion-games-in-ancient-egypt.

Mark, Joshua J. "Ur." *Ancient History Encyclopedia,* April 28, 2011, www.ancient.eu/ur.

Marshall, Rick. "Coincidence or Curse? Looking Back on Madden's Troubled Cover Athletes." *Digital Trends,* August 23, 2014, www.digitaltrends.com/gaming/the-madden-curse.

Moghadasi, Abdorreza Naser. "The Burnt City and the Evolution of the Concept of 'Probability' in the Human Brain." *Iranian Journal of Public Health* 44, no. 9 (September 2015).

Mugane, John M. *The Story of Swahili*. Athens, OH: Ohio University Press, 2015.

Oware Society, The. "History." The Oware Society, n.d., www.oware.org/history.asp.

Piccione, Peter A. "In Search of the Meaning of Senet." *Archaeology,* July/August 1980, www.gamesmuseum.uwaterloo.ca/Archives/Piccione/index.html.

———. "The Egyptian Game of Senet and the Migration of the Soul." In *Ancient Board Games In Perspective,* edited by Irving Finkel. London: British Museum Press, 2007.

Schädler, Ulrich. "Mancala in Roman Asia Minor?" *Board Game Studies* 1 (1998).

Shafer, Glenn. "The Early Development of Mathematical Probability." In *Companion Encyclopedia of History and Philosophy of the Mathematical Sciences,* edited by I. Grattan-Guinness. London: Routledge, 1993.

Shermer, Michael. "Patternicity: Finding Meaningful Patterns in Meaningful Noise." *Scientific American,* December 1, 2008, www.scientificamerican.com/article/patternicity-finding-meaningful-patterns.

Townshend, Philip. "Games of Strategy: A New Look at Correlates and Cross-Cultural Methods." Brian Sutton-Smith Archives, The Strong National Museum of Play, Rochester, NY.

Zorich, John N., Jr. "The Prehistory of Probability." Discussion paper presented at Statistics Group of the Santa Clara Valley tomb of ASQ, November 8, 2000.

### 2. Chess: The "Mad Queen's Game"

Batgirl. "Café de la Régence." Chess.com (blog), July 17, 2011, www.chess.com/blog/batgirl/cafeacute-de-la-reacutegence.

Beccia, Carlyn. "Raucous Royal of the Month, Caterina Sforza: Daughter of Perdition." *The Raucous Royals* (blog), May 9, 2010, blog.raucousroyals.com/2010/05/raucous-royal-of-month-caterina-sforza.html.

Bell, R. C. *Board and Table Games from Many Civilizations*. New York: Dover Publications, 1979.

Carr, Raymond. *Spain: A History*. Oxford: Oxford University Press, 2000.

Chaudhuri, Dola. "Gupta Empire." *Ancient History Encyclopedia,* October 30, 2015, www.ancient.eu/Gupta_Empire.

Collins, Mortimer. *Frances, Vol. III*. London: Hurst and Blackett Publishers, 1874.

Dewey, Alan, and Milissa Ellison. "Design of the Staunton Chess Set." *ChessSpy,* www.chessspy.com/articles/Staunton%20Chess%20Set%20Design.pdf.

Glonnegger, Erwin. *Das Spiele-Buch*. Uehlfeld, Germany: Drei Magier Verlag, 1999.

Inglis, Lucy. "Old Slaughter's Coffee House." *Georgian London,* September 22, 2009, georgianlondon.com/post/49464383370/old-slaughters-coffee-house.

Keene, Raymond. "The Islamic World Has Always Had a Chequered Relationship with Chess." *The Spectator,* January 22, 2016, blogs.spectator.co.uk/2016/01/the-islamic-world-has-always-had-a-chequered-relationship-with-chess.

Murray, H. J. R. *A History of Chess*. London: Oxford University Press, 1913.

Reider, Norman. "Chess, Oedipus and the Mater Dolorosa." *Psychoanalysis and the Psychoanalytic Review* 47, no. 2 (summer 1960).

Remus, Horst. "The Origins of Chess and the Silk Road." *The Silk Road Foundation Newsletter* 1, no. 1, January 15, 2003, www.silkroadfoundation.org/newsletter/volumeonenumberone/origin.html.

Shaheen, Kareem. "Chess Forbidden in Islam, Rules Saudi Mufti, But Issue Not Black and White." *Guardian,* January 21, 2016, www.theguardian.com/world/2016/jan/21/chess-forbidden-in-islam-rules-saudi-arabia-grand-mufti.

Shenk, David. *The Immortal Game: A History of Chess*. London: Souvenir Press, 2006.

Stamp, Jimmy. "How the Chess Set Got Its Look and Feel." *Smithsonian,* April 3, 2013, www.smithsonianmag.com/arts-culture/how-the-chess-set-got-its-look-and-feel-14299092/?no-ist.

Tomlinson, Charles. *Amusements in Chess*. London: John W. Parker, 1845.

Wall, Bill. "The Staunton Chessmen." *ChessManiac,* April 10, 2013, www.chessmaniac.com/the-staunton-chessmen.

Williams, Gareth. *Master Pieces: The Architecture of Chess*. London: Quintet Publishing, 2000.

Yalom, Marilyn. *Birth of the Chess Queen: A History*. London: Pandora Press, 2004.

### 3. Backgammon: The Favored Game of International Pace-Setters and Ancient Emperors

Avrett, Jack. "Everybody's Kind of Game." *Madison Avenue,* January 1974.

Bell, R. C. *Board and Table Games from Many Civilizations*. New York: Dover Publications, 1979.

Benchley, Robert C. "I Spy Backgammon." *Detroit Athletic Club News,* December 1930. In *The Athletic Benchley: 105 Exercises from the Detroit Athletic Club News* by Robert C. Benchley and edited by Thomas J. Saunders. Toronto: Glendower Media, 2010.

Bradshaw, Jon. "Backgammon." *Harper's Magazine,* June 1972.

Brekke, Dan. "Chicagoland Cemetery Report." *Infospigot* (blog), September 2, 2011, infospigot.typepad.com/infospigot_the_chronicles/2011/09/chicagoland-cemetery-report.html.

Buckley, Christopher. "Protecting Sinatra Against the Big-Beef Story." *New York,* July 15, 1974.

Davis, Bill. "Backgammon and the Doubling Cube—Just the Facts." *Chicago Point,* www.chicagopoint.com/bgdoubling.html.

Deyong, Lewis. "Sport and Leisure Europe: 1974," *Madison Avenue,* January 1974.

———. *Playboy's Book of Backgammon*. New York: Playboy Press, 1977.

———. "From Sumer to Monte Carlo Backgammon." *Society* (spring/summer 1982).

Driver, Mark. "A History of Backgammon." *Backgammon Galore,* November 2000, www.bkgm.com/articles/GOL/Nov00/mark.htm.

du Coeur, Justin. "Game Report: Irish, and Early Backgammon." March 29, 1997, jducoeur.org/game-hist/game-recon-irish.html.

Glonnegger, Erwin. *Das Spiele-Buch.* Uehlfed, Germany: Drei Magier Verlag, 1999.

Hevesi, Dennis. "Tim Holland, Backgammon Master, Dies at 79." *New York Times,* March 17, 2010, www.nytimes.com/2010/03/17/us/17holland.html?_r=0.

"History of Backgammon" *MoneyGaming,* n.d., www.moneygaming.com/skillSchool/backgammon/history.shtml.

Jacoby, Oswald, and John R. Crawford. *The Backgammon Book.* New York: Viking Press, 1970.

Kerr, Dale. "World Championships of Backgammon." *Backgammon Galore!* September 2007, www.bkgm.com/articles/Kerr/WorldChampionships.

Maxa, Rudy. "High Rolling in Monte Carlo." *GQ,* July 1984.

"Money Game." *Newsweek,* November 20, 1972.

Obolensky, Prince Alexis, and Ted James. *Backgammon: The Action Game.* New York: Collier Books, 1969.

Obolensky, Valerian. *Russians in Exile: The History of a Diaspora.* 1993, russians.bellevueholidayrentals.com/dias1.html.

Perrin, Bernadotte. "Plutarch, Life of Artaxerxes." *Livius.org,* 2007, www.livius.org/pi-pm/plutarch/plutarch_artaxerxes_1.html.

Robinson, Jeffrey. "Big Money at the Mecca of Backgammon." *International Herald Tribune,* July 8, 1977.

Rosin, Mark. "Adult Games/The Entertainment Is at Home." *Bazaar,* January 1973.

Shrake, Edwin. "Everyone for Backgammon." *Sports Illustrated,* May 6, 1964, www.si.com/vault/1964/05/04/606810/everyone-for-backgammon.

Silverman, David. "Largest Tax-Evasion Case in Area History." *Chicago Tribune,* August 18, 1993.

"The Bald Facts About Ringo . . . Hair Today and None Tomorrow," *Daily Mail,* July 19, 1976.

"The Game of Swedish Tables." Vasamuseets Brädspelsvänner, February 26, 2003, www.vasamuseet.se/globalassets/vasamuseet/dokument/om/bradspel_eng_regelr.pdf.

"The Money Game," *Time,* February 19, 1972.

Tylor, Edward B. "Backgammon Among the Aztecs." Brian Sutton-Smith Archives, The Strong National Museum of Play, Rochester, NY.

"World Backgammon Championship 1967 to 1979: Results and Some Historical Notes." *Gammon Press,* March 9, 2015, thegammonpress.com/world-backgammon-championship-1967-to-1979-results-and-some-historical-notes.

#### 4. The Game of Life: A Journey to the Uniquely American Day of Reckoning

"'All the Ladies Like Whiskers!' Revealed, the Girl, 11, Who Convinced Lincoln to Grow His Iconic Beard." *Mail Online,* November 30, 2012, www.dailymail.co.uk/news/article-2240765/Grace-Bedell-Abraham-Lincoln-grew-beard-girl-11-wrote-said-ladies-like-whiskers.html.

Angiolillo, Joseph. "W. & S. B. Ives, Part II [The Mansion of Happiness]." BoardGameGeek (blog), August 18, 2011, boardgamegeek.com/thread/687986/w-s-b-ives-part-ii-mansion-happiness-joseph-angiol.

Bradley, Milton. "The Checkered Game of Life." Board game rules, Milton Bradley Company, 1860.

———. "Games and Amusements as Benefits and Blessings in the Home." *Good Housekeeping,* January 1896.

Bradley, Milton, to Stewart Culin, April 10, 1893, Culin Archival Collection, Brooklyn Museum, Brooklyn, NY.

Cohen, Alma, and Rajeev Dehejia. "The Effect of Automobile Insurance and Accident Liability Laws on Traffic Fatalities," *Journal of Law and Economics* 47, no. 2 (2004).

*Economist, The.* "The End of Jobs for Life?" *Economist,* February 19, 1998, www.economist.com/node/604599.

Farber, Henry S. "Employment Insecurity: The Decline of Worker-Firm Attachment in the United States." CEPS Working Paper No. 79, January 2008, www.princeton.edu/ceps/workingpapers/172farber.pdf.

Finefield, Kristi. "A Look Back at Board Games." *Library of Congress* (blog), April 3, 2014, blogs.loc.gov/picturethis/2014/04/a-look-back-at-board-games.

Glonnegger, Erwin. *Das Spiele-Buch.* Uehlfeld, Germany: Drei Magier Verlag, 1999.

Hofer, Margaret K. *The Games We Played: The Golden Age of Board & Table Games.* New York: Princeton Architectural Press.

Milton Bradley Company. "The Game of Life." Board game rules, 1960.

National Radiation Instrument Catalog. "The Uranium Rush—1949." National Radiation Instrument Catalog, n.d., national-radiation-instrument-catalog.com/new_page_14.htm.

Neumark, David, Daniel Polsky, and Daniel Hansen. "Has Job Stability Declined Yet? New Evidence for the 1990's." *National Bureau of Economic Research,* December 1997, www.nber.org/papers/w6330.pdf.

Orbanes, Philip E. *The Game Makers: The Story of Parker Brothers, from Tiddledy Winks to Trivial Pursuit.* Boston: Harvard Business School Press, 2004.

Shea, James J., Sr., and Charles Mercer. *It's All in the Game.* New York: G.P. Putnam's Sons, 1960.

Stashower, David. "The Truth About Lincoln's Beard." *The History Reader,* April 1, 2014, www.thehistoryreader.com/modern-history/truth-lincolns-beard.

"Trends in Political Values and Core Attitudes: 1987–2002." Pew Research Center, March 22, 2007, www.people-press.org/2007/03/22/trends-in-political-values-and-core-attitudes-1987-2007.

W. & S. B. Ives. "The Mansion of Happiness." Board game rules, 1843.

Wagner, David. *The Poorhouse: America's Forgotten Institution.* Lanham, MD: Rowman & Littlefield, 2005.

Walsh, Tim. *Timeless Toys: Classic Toys and the Playmakers Who Created Them.* Kansas City: Andrews McMeel Publishing, 2005.

Zoellner, Tom. *Uranium: War, Energy and the Rock That Shaped the World.* London: Viking, 2009.

## 5. The Forgotten Message of Monopoly

Andrews, Frank Emerson. *Corporation Giving.* New Brunswick, NJ: Transaction Publishers, 1993.

Barton, Robert, to Victor H. Watson, August 19, 1957, John Waddington's PLC Collection, West Yorkshire Archive Service, Leeds, UK.

Brady, Maxine. "Everything You Never Knew About Monopoly," *Pastimes,* October 1974.

———. *The Monopoly Book: Strategy and Tactics of the World's Most Popular Game.* London: Robert Hale, 1974.

Brogan, Hugh. *The Penguin History of the USA.* London: Penguin, 1999.

Center for Community Solutions, The. "Teamwork For a Better Cleveland 1913–2013." 2013, www.communitysolutions.com/assets/docs/CCS_General/2014_ccs_centennial_publication_update_062314.pdf.

Chevan, Albert. "The Growth of Home Ownership: 1940–1980," *Demography* 26, no. 2 (May 1989).

Darrow, Charles B., to Robert Barton, March 21, 1935, Philip E. Orbanes Archives, The Strong National Museum of Play, Rochester, NY.

George, Henry. *Progress and Poverty.* New York: E.P. Dutton & Company, 1879.

George, Henry, and Kenneth C. Wenzer. *An Anthology of Henry George's Thought, Vol. 1.* Rochester, NY: University of Rochester Press, 1997.

Hilton, George W., and John F. Due. *The Electric Interurban Railways in America.* Stanford: Stanford University Press, 2000.

Ignatius, Adi. "Capitalist Game Big Hit With Socialists," unknown publication, October 6, 1988, Philip E. Orbanes Archive, The Strong National Museum of Play, Rochester, NY.

Magie, Elizabeth. "The Landlord's Game." *The Single Tax Review,* autumn 1902, lvtfan.typepad.com/lvtfans_blog/2011/01/lizzie-magie-1902-commentary-the-landlords-game.html.

Marx, Karl, to Friedrich Adolph Sorge, June 20, 1881, www.marxists.org/archive/marx/works/1881/letters/81_06_20.htm.

"Monopoly Sales 1935–1974." N.d., Philip E. Orbanes Archive, The Strong National Museum of Play, Rochester, NY.

"Monopoly Sales by Waddingtons in UK." N.d., John Waddington's PLC Collection, West Yorkshire Archive Service, Leeds, UK.

Morgan, Robin. "Hot Property." *Yorkshire Post,* January 14, 1985.

Orbanes, Philip E. *The Game Makers: The Story of Parker Brothers, from Tiddledy Winks to Trivial Pursuit.* Boston, MA: Harvard Business Press, 2004.

———. *Monopoly: The World's Most Famous Game & How It Got That Way.* Philadelphia: De Capo Press, 2006.

———. "Meet Dan Fox: The Artist Who Created 'Mr. Monopoly.'" June 4, 2013, Philip E. Orbanes Archive, The Strong National Museum of Play, Rochester, NY.

Pilon, Mary. *The Monopolists: Obsession, Fury, and the Scandal Behind the World's Favorite Board Game.* New York: Bloomsbury, 2015.

"Real Riches From Monopoly Game," *Times,* April 21, 1959.

"Re: Parker Brothers, Inc." Unsigned memo, May 2, 1960, John Waddington's PLC Collection, West Yorkshire Archive Service, Leeds, UK.

Simon, Roger D. "Philadelphia and the Great Depression." *The Encyclopedia of Greater Philadelphia,* 2013, philadelphiaencyclopedia.org/archive/great-depression.

Stiles, T. J. "Robber Barons or Captains of Industry?" *History Now,* n.d., www.gilderlehrman.org/history-by-era/gilded-age/essays/robber-barons-or-captains-industry.

"The Great Strike," *Harper's Weekly,* August 11, 1877, www.catskillarchive.com/rrextra/sk7711.html.

Vandegrift, Juliana. "Mr Victor Watson and Mr Colin Linn Interviews," Museum of Childhood, 2012.

Waddingtons Games. *50 Years of Monopoly.* Leeds, 1985, John Waddington's PLC Collection, West Yorkshire Archive Service, Leeds, UK.

Walsh, Tim. *Timeless Toys: Classic Toys and the Playmarkers Who Created Them.* Kansas City: Andrews McMeel Publishing, 2005.

### 6. From Kriegsspiel to Risk: Blood-Soaked and World-Shaping Play

Campion, Martin, and Steven Patrick. "The History of Wargaming." *S&T Magazine,* July 1972.

Donovan, Tristan. *Replay: The History of Video Games.* Lewes, UK: Yellow Ant, 2010.

Handcock, Peter A., et al (eds.) *Human Factors in Simulation and Training.* Boca Raton, FL: CRC Press, 2009.

Johnson, Moira. "It's Only a Game—Or Is It?" *New West,* August 25, 1980.

Lamorisse, Albert-Emmanuel. Perfectionnements apportés aux jeux de société. French patent 1,101,756, filed March 23, 1954, and issued October 11, 1955.

Leeson, Bill. "Origins of the Kriegsspiel." *Kriegsspiel News,* n.d., www.kriegsspiel.org.uk/index.php/articles/origins-history-of-kriegsspiel/3-origins-of-the-kriegsspiel.

Neville, Peter. *Russia: A Complete History in One Volume.* Moreton-in-Marsh, UK: Windrush Press, 2000.

Orbanes, Philip E. *The Game Makers: The Story of Parker Brothers, from Tiddledy Winks to Trivial Pursuit.* Boston, MA: Harvard Business School Press, 2004.

Prange, Gordon W., with Donald M. Goldstein and Katherine V. Dillon. *At Dawn We Slept: The Untold Story of Pearl Harbor.* New York: Penguin, 1981.

Reider, Norman. "Chess, Oedipus and the Mater Dolorosa." *Psychoanalysis and the Psychoanalytic Review* 47, no. 2 (summer 1960).

Roberts, Charles S. "Charles S. Roberts: In His Own Words" 1983, www.alanemrich.com/CSR_pages/Articles/CSRspeaks.htm.

Shapiro, Dave. "A Conversation with Roberto Convenevole." *BoardGameGeek* (blog), March 4, 2010, boardgamegeek.com/thread/501810/conversation-roberto-convenevole.

Vego, Milan. "German War Gaming." *Naval War College Review* 65, no. 4 (autumn 2012).

von Hilgers, Philipp. *War Games: A History of War on Paper.* Translated by Ross Benjamin. Cambridge, MA: MIT Press, 2012.

von Reisswitz, B. *Kriegsspiel.* 1824, translated by Bill Leeson, Hemel Hempsted, UK: 1989.

Wells, H. G. *Little Wars.* London: Frank Palmer, 1913.

Wintjes, Jorit. "Europe's Earliest Kreigsspiel?" *British Journal for Military History* 2, no. 1 (November 2015).

Wolff, David, et al (eds.) *The Russo-Japanese War in Global Perspective: World War Zero, Vol. 2.* Leiden, Netherlands: Brill, 2007.

### 7. I Spy

Barden, Leonard. "Obituary: Bobby Fischer." *Guardian,* January 18, 2008, www.theguardian.com/obituaries/story/0,2243089,00.html.

Botvinnik, Mikhail. *One Hundred Selected Games*. Translated by Stephen Garry. New York: MacGibbon & Kee, 1951.

Chun, Rene. "Bobby Fischer's Pathetic Endgame." *Atlantic*, December 2002.

Donlan, Christian. "Inside Monopoly's Secret War Against the Third Reich." *Eurogamer*, January 12, 2014, www.eurogamer.net/articles/2014-01-12-inside-monopolys-secret-war-against-the-third-reich.

Edmonds, David, and John Eidinow. *Bobby Fischer Goes to War*. London: Faber and Faber, 2004.

Evans, J. G., to Victor Watson, January 16, 1985, John Waddington PLC Collection, West Yorkshire Archive Service, Leeds, UK.

Friedman, David. "Paul Keres." *Ohio Chess Connection*, November/December 2008.

Ginzburg, Ralph. "Portrait of a Genius as a Young Chess Master." *Harper's Magazine*, January 1962.

Gulko, Boris, Vladimir Popov, Yuri Felshtinsky, and Viktor Kortschnoi. *The KGB Plays Chess*. Milford, CT: Russell Enterprises, 2010.

Hall, Debbie. "Wall Tiles and Free Parking: Escape and Evasion Maps of World War II." *Antique Map Magazine* 4, n.d., www.mapforum.com/04/april.htm.

*Historical Record of IS9*. WO 208/3242, The National Archives, London.

*Historical Record of MI9*. WO 208/3242, The National Archives, London.

Lawton, W. T. "Bill," to Victor Watson, January 15, 1985, John Waddington PLC Collection, West Yorkshire Archive Service, Leeds, UK.

Morgan, Robin. "Hot Property." *Yorkshire Post*, January 14, 1985.

No. 9 Intelligence School. "Lecture Notes On Secret Code Letter Writing." WO 208/3242, The National Archives, London.

Orbanes, Philip E. *Monopoly: The World's Most Famous Game & How It Got That Way*. Philadelphia, PA: De Capo Press, 2006.

———. "Monopoly, Code Users, and POWs In World War II." *Association of Game & Puzzle Collectors Quarterly*, spring 2015.

Pilon, Mary. *The Monopolists: Obsession, Fury, and the Scandal Behind the World's Favorite Board Game*. New York, NY: Bloomsbury, 2015.

"Postcards Containing Cold War Spy Messages Unearthed." *Telegraph*, July 24, 2009, www.telegraph.co.uk/news/uknews/5895953/Postcards-containing-Cold-War-spy-messages-unearthed.html.

Robson, J. T., to Victor Watson, January 22, 1985, John Waddington PLC Collection, West Yorkshire Archive Service, Leeds, UK.

Rosin, Mark. "Adult Games/The Entertainment Is at Home." *Bazaar*, January 1973.

Shenk, David. *The Immortal Game: A History of Chess*. London: Souvenir Press, 2006.

Souvarine, Boris. *Stalin: A Critical Survey of Bolshevism*. Translated by C. L. R. James. New York, NY: Alliance Book Corporation, 1939.

"Stalag XX a Thorn." WO 208/3281, The National Archives, London.

Unsigned letter to Mrs T. J. Walker, February 20, 1985, John Waddington PLC Collection, West Yorkshire Archive Service, Leeds, UK.

Watson, Norman, to Charles McConnell, May 27, 1944, John Waddington PLC Collection, West Yorkshire Archive Service, Leeds, UK.

Watson, Norman. "M.I.5 & No. 40, Wakefield Road," *Waddingtons Magazine,* May 1968.

Williams, Gareth. *Master Pieces: The Architecture of Chess.* London: Quintet Publishing, 2000.

## 8. Clue's Billion-Dollar Crime Spree

Akers, Michael. The Art of Murder, n.d., www.theartofmurder.com.

Barton, Robert, to Norman Watson, May 26, 1948, John Waddington PLC Collection, West Yorkshire Archive Service, Leeds, UK.

Barton, Robert, to Norman Watson, June 11, 1948, John Waddington PLC Collection, West Yorkshire Archive Service, Leeds, UK.

Barton, Robert, to Norman Watson, August 30, 1948, John Waddington PLC Collection, West Yorkshire Archive Service, Leeds, UK.

Barton, Robert, to Norman Watson, June 29, 1949, John Waddington PLC Collection, West Yorkshire Archive Service, Leeds, UK.

Barton, Robert, to Norman Watson, June 30, 1949, John Waddington PLC Collection, West Yorkshire Archive Service, Leeds, UK.

Barton, Robert, to Norman Watson, March 10, 1950, John Waddington PLC Collection, West Yorkshire Archive Service, Leeds, UK.

Barton, Robert, to Norman Watson, April 21, 1954, John Waddington PLC Collection, West Yorkshire Archive Service, Leeds, UK.

Cauterucci, Christina. "Clue Changes Out Mrs. White, a housekeeper, for Dr. Orchid, a female scientist." *Slate*, July 8, 2016, www.slate.com/blogs/xx_factor/2016/07/08/_clue_changes_out_mrs_white_a_housekeeper_for_dr_orchid_a_female_scientist.html.

"Cluedo Agreement History." No date or author, John Waddington PLC Collection, West Yorkshire Archive Service, Leeds, UK.

Foster, Jonathan. *The Story of Cluedo: How Anthony Pratt Invented the Game of Murder Mystery.* York, UK: York Publishing, 2013.

Hoban, Phoebe. "Revenge of the Video Game." *New York,* April 28, 1986.

"Inventor of Clue Dies Amid Mystery Worthy of His Game." *Sun-Journal* (Lewiston, ME), December 2, 1996.

Kindred, Michael. *Once Upon a Game . . . My Precarious Career as a Games Inventor.* Dartford, Kent: Penuma Springs Publishing, 2013.

Orbanes, Philip E. *The Game Makers: The Story of Parker Brothers, from Tiddledy Winks to Trivial Pursuit.* Boston: Harvard Business School Press, 2004.

Pratt, Anthony, to Norman Watson, February 27, 1967, John Waddington PLC Collection, West Yorkshire Archive Service, Leeds, UK.

Sandbrook, Dominic. *The Great British Dream Factory: The Strange History of Our National Imagination*. London: Allen Lane, 2015.

Strickler, Jeff. "Seeing 'Clue' Is Like Playing the Game." *Minneapolis Star and Tribune,* December 13, 1985.

Strong, Roy. *Visions of England or Why We Still Dream of a Place in the Country*. London: Vintage Books, 2011.

Summerscale, Kate. "Jack Mustard, In the Spa, With a Baseball Bat." *Guardian,* December 20, 2008, www.theguardian.com/lifeandstyle/2008/dec/20/cluedo-new-rebrand-family.

Treneman, Ann. "Mr Pratt, In the Old People's Home, With an Empty Pocket." *Independent,* November 12, 1998, www.independent.co.uk/arts-entertainment/mr-pratt-in-the-old-peoples-home-with-an-empty-pocket-1184258.html.

Utton, Dominic. "The Forgotten Mr Cluedo." *Express,* August 17, 2009, www.express.co.uk/expressyourself/120990/The-forgotten-Mr-Cluedo.

Walsh, Tim. *Timeless Toys: Classic Toys and the Playmakers Who Created Them*. Kansas City: Andrews McMeel Publishing, 2005.

Watson, John. "Cluedo." September 22, 1975, John Waddington PLC Collection, West Yorkshire Archive Service, Leeds, UK.

Watson, Norman, to Anthony Pratt, August 31, 1945, John Waddington PLC Collection, West Yorkshire Archive Service, Leeds, UK.

Watson, Norman, to Robert Barton, January 21, 1947, John Waddington PLC Collection, West Yorkshire Archive Service, Leeds, UK.

Watson, Norman, to Robert Barton, June 8, 1948, John Waddington PLC Collection, West Yorkshire Archive Service, Leeds, UK.

Watson, Norman, to G. G. Bull, February 22, 1951, John Waddington PLC Collection, West Yorkshire Archive Service, Leeds, UK.

Watson, Victor H., to Mr. B. R. Watson, Mr. S. Boyd and Mrs. J. Bentley. "My Comments on the New Cluedo." Memo, September 2, 1985, John Waddington PLC Collection, West Yorkshire Archive Service, Leeds, UK.

Whitehill, Bruce. *The Story of Cluedo & Clue-Part 1,* n.d., unpublished.

Worsley, Lucy. *A Very British Murder.* London: BBC Books, 2013.

### 9. Scrabble: Words Without Meaning

"And Scrabble, Too." *Madison Avenue,* January 1974.

Bethea, Charles. "The Battle Over Scrabble's Dictionaries." *New Yorker,* August 4, 2015, www.newyorker.com/news/sporting-scene/the-battle-over-scrabbles-dictionaries.

Burns, Russell W. *Communications: An International History of the Formative Years.* London: The Institution of Electrical Engineers, 2004.

Elliot, George. "Brief History of Crossword Puzzles." American Crossword Puzzle Tournament website, www.crosswordtournament.com/more/wynne.html.

Ember, Sydney. "For a Bereft Street Corner in Queens, a Red-Letter Day." *New York Times,* July 15, 2011, www.nytimes.com/2011/07/16/nyregion/sign-in-queens-marking-birthplace-of-scrabble-is-coming-back.html?_r=1.

Fatsis, Stefan. *Word Freak.* London: Penguin Books, 2001.

Jaeger, Philip Edward. *Cedar Grove.* Charleston, SC: Arcadia Publishing, 2000.

Santoso, Alex. "The Origin of the Crossworld Puzzle." *Neatorama,* March 31, 2008, www.neatorama.com/2008/03/31/the-origin-of-the-crossword-puzzle.

Shaer, Matthew. "How Crossword Inventor Arthur Wynne Designed His First Puzzle." *Christian Science Monitor,* December 22, 2013, www.csmonitor.com/Technology/2013/1222/How-crossword-inventor-Arthur-Wynne-designed-his-first-puzzle.

Spear, Francis. "History." Spear's Games Archives website, www.spearsgamesarchive.co.uk/content-index/history.

Wallace, Robert. "Little Business in the Country." *Life,* December 14, 1953.

Walsh, Tim. *Timeless Toys: Classic Toys and the Playmakers Who Created Them.* Kansas City: Andrews McMeel Publishing, 2005.

Wepman, David. "Butts, Alfred Mosher." *American National Biography Online,* February 2000, www.anb.org/articles/20/20-01922-print.html.

Williams Jr., John D. *Word Nerd: Dispatches from the Games, Grammar, and Geek Underground.* New York, NY: Liveright Publishing, 2015.

Willsher, Kim. "The French Scrabble Champion Who Doesn't Speak French." *Guardian,* July 21, 2015, www.theguardian.com/lifeandstyle/2015/jul/21/new-french-scrabble-champion-nigel-richards-doesnt-speak-french.

Worley, Sam. "The Puzzler and the Puzzled." *Chicago Reader,* March 28, 2012, www.chicagoreader.com/chicago/the-puzzler-and-the-puzzled/Content?oid=5932279.

### 10. Plastic Fantastic: Mouse Trap, Operation, and the Willy Wonka of Toys

"American Toy Designer Is Modern 'Pied Piper.'" *Schenectady* (New York) *Gazette,* November 20, 1968.

Anderson, Robert. "Chicago's Toy Genius." *Chicago Sunday Tribune Magazine,* November 5, 1961.

Cushman, Aaron D. *A Passion For Winning: Fifty Years of Promoting Legendary People and Products.* Pittsburgh, PA: Lighthouse Point Press, 2004.

Erickson, Erick. "Recollections of Working at the Marvin Glass Studio." N.d., www.marvinglass.com/index.html.

"Games: Past Go and Still Accelerating." *Toys and Novelties,* July 1970.

"He's King of the Mechanical Toy Manufacturers." *Daytona Beach Sunday News Journal,* February 28, 1954.

Hix, Lisa. "The Inside Scoop on Fake Barf." *Collectors Weekly,* August 23, 2011, www.collectorsweekly.com/articles/the-inside-scoop-on-the-fake-barf-industry.

Hughes, Alice. "Flowers." *Palm Beach Daily,* March 19, 1962.

"Leading Toy Designer Tops All Records With Projects." *Lawrence* (Kansas) *Journal-World* (Lawrence, KS), November 25, 1965.

Lewis, Caroline. "Old Coach House a Showplace." *Chicago Tribune,* January 5, 1964.

Moye, David. "John Spinello, Inventor of 'Operation' Game, Can't Afford Real-Life Operation (Updated)." *Huffington Post,* October 27, 2014, www.huffingtonpost.com/2014/10/27/john-spinello_n_6055174.html.

Orbanes, Philip E. *The Game Makers: The Story of Parker Brothers, From Tiddledy Winks to Trivial Pursuit.* Boston, MA: Harvard Business School Press.

"A Playboy Pad: Swinging in Suburbia." *Playboy,* May 1970.

"Marvin Glass, Designer, Dead At 59; Industry Dean." *Playthings,* February 1974.

"Secrecy Surrounds Toy Making." *Rome News-Tribune* (Rome, GA), April 25, 1971.

"Teaching Toys in Giant Strides." *Life,* December 12, 1960.

"Toy Designer a Child at Heart." *Milwaukee Journal,* November 26, 1972.

"Toy King Worries About Spies." *Sarasota* (Florida) *Herald-Tribune,* April 25, 1971.

"Voices," *Life,* November 24, 1961.

Walsh, Tim. *Timeless Toys: Classic Toys and the Playmakers Who Created Them.* Kansas City: Andrews McMeel Publishing, 2005.

Wyden, Peter. "Troubled King of Toys." *Saturday Evening Post,* March 5, 1960.

### 11. Sex in a Box

Associated Press. "Dr. Ruth of 'Good Sex' Inspires New Board Game." *Nashua* (New Hampshire) *Telegraph,* August 21, 1985.

Cohen, Nancy L. "How the Sexual Revolution Changed America Forever." *Alternet,* February 5, 2012, www.alternet.org/story/153969/how_the_sexual_revolution_changed_america_forever.

Guyer, Reyn. "Twister History," unpublished, 1994.

Lyall, Sarah. "Revising 'Sex' for the 21st Century." *New York Times,* December 17, 2008, www.nytimes.com/2008/12/18/fashion/18joy.html.

Mikkelson, Barbara. "Getting Harrier'd Away." *Snopes,* May 5, 2011, www.snopes.com/business/deals/pepsijet.asp.

"New Board Game Doesn't Trivialize Sex," *The Day* (New London, CT), January 11, 1987.

Parker, Alex. "Crossing Over: U.K.-Based Creative Conceptions Expands into the U.S." *Xbiz,* April 9, 2014, www.xbiz.com/articles/177432/creative+conceptions.

Rottenburg, Dan. "Can You Come Over For 'Mob Strategy' Tonight—Or Will It Be a Game of 'Adultery'?" *Today's Health,* August 1971.

Sweeney, Brigid. "Sears—Where America Shopped." *Crain's Chicago Business,* April 21, 2012, www.chicagobusiness.com/article/20120421/ISSUE01/304219970/sears-where-america-shopped.

Timberg, Bernard M., and Robert J. Erler. *Television Talk: A History of the TV Talk Show.* Austin, TX: University of Texas Press, 2002.

Twenge, Jean M., Ryne A. Sherman, and Brooke E. Wells. "Changes in American Adults' Sexual Behavior and Attitudes, 1972–2012." *Archives of Sexual Behavior* 44, no. 8 (November 2015).

Walsh, Tim. *Timeless Toys: Classic Toys and the Playmakers Who Created Them.* Kansas City: Andrews McMeel, 2005.

## 12. Mind Games: Exploring Brains with Board Games

Crestbook. "KC-Conference With Judit Polgar." *Crestbook,* January 12, 2012, www.crestbook.com/node/1668.

Dingfelder, Sadie F. "Linear or Logarithmic?" *Monitor on Psychology* 36, no. 10 (November 2005).

Flora, Carlin. "The Grandmaster Experiment." *Psychology Today,* July 1, 2005, www.psychologytoday.com/articles/200507/the-grandmaster-experiment.

Furedi, Frank. *Therapy Culture: Cultivating Vulnerability in an Uncertain Age.* London: Routledge, 2008.

Gobet, Fernand, Alex de Voogt, and Jean Retschitzki. *Moves in Mind: The Psychology of Board Games.* Hove, UK: Psychology Press, 2004.

Hearst, Eliot, and John Knott. *Blindfold Chess: History, Psychology, Techniques, Champions, World Records, and Important Games.* Jefferson, NC: McFarland & Company, 2009.

Hoare, Rose. "Judit Polgár, the Chess Prodigy Who Beat Men at Their Own Game." *CNN,* September 3, 2012, edition.cnn.com/2012/08/30/world/europe/judit-polgar/index.html.

Jarvik, Elaine. "Rhea Zakich Discovers Key to Communication," *Deseret News* (Salt Lake City, UT), November 18, 1985.

Lundstrom, Harold. "Father Of 3 Prodigies Says Chess Genius Can Be Taught," *Deseret News* (Salt Lake City, UT), December 25, 1992, www.deseretnews.com/article/266378/FATHER-OF-3-PRODIGIES-SAYS-CHESS-GENIUS-CAN-BE-TAUGHT.html?pg=all.

McDonald, Patrick S. *The Benefits of Chess in Education: A Collection of Studies and Papers on Chess and Education.* N.d., www.psmcd.net/otherfiles/BenefitsOfChessInEdScreen2.pdf.

"Nurtured to Be Geniuses, Hungary's Polgar Sisters Put Winning Moves on Chess Masters." *People,* May 1987, www.people.com/people/article/0,20096193,00.html.

Polgár, Judit. "Judit Polgar's Official Statement on Her Retirement." N.d., juditpolgar.com/en/node/230.

Rassin-Gutman, Diego. *Chess Metaphors: Artificial Intelligence and the Human Mind.* Translated by Deborah Klosky. Cambridge, MA: MIT Press, 2009.

Rottenberg, Dan. "Can You Come Over For 'Mob Strategy' Tonight—Or Will It Be a Game of 'Adultery'?" *Today's Health,* August 1971.

Shenk, David. *The Immortal Game: A History of Chess.* London: Souvenir Press, 2006.

Smith, Beverly J. "The Grown-Up Game Craze: Can Reality Be Put in a Box," *Philadelphia Inquirer Magazine,* January 9, 1972.

Vedantam, Shankar. "Mind Games May Trump Alzheimer's." *Washington Post,* June 19, 2003.

### 13. Rise of the Machines: Games That Train Synthetic Brains

Akpan, Nsikan. "How a Computer Program Became Champion of the World's Trickest Board Game." *PBS Newshour,* January 27, 2016, www.pbs.org/newshour/bb/how-a-computer-program-became-champion-of-the-worlds-trickiest-board-game.

"Almanac: Kasparov Vs. Deep Blue." *CBS News,* May 3, 2015. www.cbsnews.com/news/almanac-kasparov-vs-deep-blue.

Amhed, Kamal. "Google's Demis Hassabis—Misuse of Artificial Intelligence 'Could Do Harm.'" *BBC News,* September 16, 2015, www.bbc.co.uk/news/business-34266425.

Asakawa, Naoki. "Flaws in AI Seen Despite AlphaGo Victory." *Nikkei Asian Review,* March 20, 2016, asia.nikkei.com/Tech-Science/Tech/Flaws-in-AI-seen-despite-AlphaGo-victory?page=1.

Bell, R. C. *Board and Table Games From Many Civilizations.* New York: Dover Publications, 1979.

Bernstein, Alex, and Michael de V. Roberts. "Computer V. Chess-Player." *Scientific American,* June 1958.

Clark, Jack. "Google Cuts Its Giant Electricity Bill With DeepMind-powered AI." *Bloomberg,* July 19, 2016, www.bloomberg.com/news/articles/2016-07-19/google-cuts-its-giant-electricity-bill-with-deepmind-powered-ai.

Cookson, Clive. "Google Computer Triumphs in Complex Board Game Battle." *Financial Times,* January 27, 2016, www.ft.com/content/b8e38a28-c4fa-11e5-b3b1-7b2481276e45.

"Deep Blue." IBM, n.d., www03.ibm.com/ibm/history/ibm100/us/en/icons/deepblue.

Donovan, Tristan. *Replay: The History of Video Games.* Lewes, UK: Yellow Ant, 2010.

Finley, Klint. "Did a Computer Bug Help Deep Blue Beat Kasparov?" *Wired,* September 28, 2012, www.wired.com/2012/09/deep-blue-computer-bug.

Gelly, Sylvain, et al. "The Grand Challenge of Computer Go: Monte Carlo Tree Search and Extensions." *Communications of the ACM* 55, no. 3 (March 2012).

Gobet, Fernand, Alex de Voogt, and Jean Retschitzki. *Moves in Mind: The Psychology of Board Games*. Hove, UK: Psychology Press, 2004.

James, Mike. "AlphaGo Has Lost a Game—Score Stands at 3-1." *I Programmer,* March 13, 2016, www.i-programmer.info/news/105-artificial-intelligence/9518-alpha-go-v-best-human-its-1-0.html.

Johnson, George. "To Test a Powerful Computer, Play an Ancient Game," *New York Times,* July 29, 1997.

Kahng, Jee Heun, and Se Young Lee. "Google Artificial Intelligence Program Beats S. Korean Go Pro With 4–1 Score." *Reuters,* March 15, 2016, www.reuters.com/article/us-science-intelligence-go-idUSKCN0WH0XJ.

Levinovitz, Alan. "The Mystery of Go, the Ancient Game That Computers Still Can't Win." *Wired,* December 5, 2014, www.wired.com/2014/05/the-world-of-computer-go.

Mack, Eric. "Elon Musk: 'We Are Summoning the Demon' With Artificial Intelligence." *CNET,* October 26, 2014, www.cnet.com/uk/news/elon-musk-we-are-summoning-the-demon-with-artificial-intelligence.

Mackenzie, Dana. "Update: Why This Week's Man-Versus-Machine Go Match Doesn't Matter (and What Does)." *Science,* March 15, 2016, www.sciencemag.org/news/2016/03/update-why-week-s-man-versus-machine-go-match-doesn-t-matter-and-what-does.

Marcus, Gary. "Go, Marvin Minsky, and the Chasm that AI Hasn't Yet Crossed." *Backchannel,* January 28, 2016, backchannel.com/has-deepmind-really-passed-go-adc85e256bec#.u1ub2j82n.

Marr, Andrew. *A History of the World*. London: Macmillan, 2012.

Metz, Cade. "Google and Facebook Race to Solve the Ancient Game of Go with AI." *Wired,* July 7, 2015. www.wired.com/2015/12/google-and-facebook-race-to-solve-the-ancient-game-of-go.

Murray, H. J. R. *A History of Chess*. London: Oxford University Press, 1913.

"Number of People Per Household in the United States from 1960 to 2015, www.statista.com/statistics/183648/average-size-of-households-in-the-us.

"Plays of Meaning." *Fields of Play* television documentary, BBC Two. London, UK: March 23, 1982.

Rasskin-Gutman, Diego. *Chess Metaphors: Artificial Intelligence and the Human Mind*. Translated by Deborah Klosky. Cambridge, MA: The MIT Press, 2009.

"Rise of the Machines." *Economist,* May 9, 2015, www.economist.com/news/briefing/21650526-artificial-intelligence-scares-peopleexcessively-so-rise-machines.

Shannon, Claude E. "Programming a Computer for Playing Chess," *Philosophical Magazine* 41 (7), no. 314 (March 1950).

Shenk, David. *The Immortal Game: A History of Chess*. London: Souvenir Press, 2006.

Shotwell, Peter. *Go! More Than a Game: Revised Edition*. North Clarendon, VT: Tuttle Publishing, 2010.

Smith, Arthur. *The Game of Go: The National Game of Japan*. New York, NY: Charles E. Tuttle Company, 1956.

Tian, Yuandong, and Yan Zhu. "Better Computer Go Player with Neural Network and Long-Term Prediction." *ArXiv*, November 19, 2015, arxiv.org/pdf/1511.06410v1.pdf.

Tran, Mark. "Go Humans: Lee Sedol Scores First Victory Against Supercomputer." *Guardian*, March 13, 2016, www.theguardian.com/world/2016/mar/13/go-humans-lee-sedol-scores-first-victory-against-supercomputer.

Weber, Peter. "Google AI Machine Beats Go Master in World's Hardest Board Game." *The Week*, March 10, 2016, theweek.com/speedreads/611808/google-ai-machine-beats-master-worlds-hardest-board-game.

Yang, Lihui, and Deming An. *Handbook of Chinese Mythology*. Santa Barbara: ABC Clio, 2005.

Zoflagharifard, Ellie. "Don't Let AI Take Our Jobs (Or Kill Us): Stephen Hawking and Elon Musk Sign Open Letter Warning of a Robot Uprising." *Mail Online*, January 12, 2015, www.dailymail.co.uk/sciencetech/article-2907069/Don-t-let-AI-jobs-kill-Stephen-Hawking-Elon-Musk-sign-open-letter-warning-robot-uprising.html#ixzz4MFoDdhVc.

## 14. Trivial Pursuit: Adults at Play

Aguiar, Mark, and Erik Hurst. "A Summary of Trends in U.S. Time Use: 1965–2005." Research paper, University of Chicago, May 2008, faculty.chicagobooth.edu/erik.hurst/research/leisure_summary_robinson_v1.pdf.

Alexander, Ron. "Trivial Pursuit Makers Ask the Right Questions, Have the Right Answers." *Lakeland* (Florida) *Ledger* January 22, 1984.

"Americans Are No. 1 in Leisure-Time Activities, Researcher Says." *Deseret News* (Salt Lake City, UT), April 9, 1990.

"And for Adults, Two Popular Games." Unknown publication, 1986, Philip E. Orbanes Archives, The Strong National Museum of Play, Rochester, NY.

Auf der Maur, Nick. "Scruples Creator Is Immersed in the Life and Works of Mysterious Author." *Montreal Gazette*, October 25, 1985.

Beamon, Bill. "Test Your Trivia I.Q." *St. Petersburg* (Florida) *Evening Independent*, May 18, 1984.

"Making Leisure Time Count." *Toledo* (Ohio) *Blade*, July 16, 1985. (Christian Science Monitor Service.)

Fialka, John J. "Inventors of a Game Score By Making Lots of People Mad." *Wall Street Journal*, 1981. Philip E. Orbanes Archives, The Strong National Museum of Play, Rochester, NY.

"The Gang That Got Away." *Inc.*, September 1988.

"The High School Dropout Who Co-Created Trivial Pursuit." *Today I Found Out*,

September 10, 2014, www.todayifoundout.com/index.php/2014/09/brief-history-trivial-pursuit.

Hollie, Pamela G. "What's New in Board Games for Adults." *New York Times,* December 15, 1985.

Jannke, Art. "All in the Game." *Boston, Inc.,* October 1986.

"Kenner Parker Update." *Toys 'n' Playthings,* March 1986.

Liegey, Paul R. "Hedonic Quality Adjustment Methods for Microwave Ovens in the U.S. CPI." Bureau of Labor Statistics, October 16, 2001, www.bls.gov/cpi/cpimwo.htm.

Long, Marion. "The Comeback of Board Games Is Breaking Video's Monopoly." Unknown publication, 1985, Philip E. Orbanes Archives, The Strong National Museum of Play, Rochester, NY.

Makow, Henry. "Rock Music's Satanic Message." HenryMakow.com, October 28, 2006, www.savethemales.ca/001799.html.

———. "9-11 Truth-Deniers Are Criminally Responsible." HenryMakow.com, September 7, 2009, www.henrymakow.com/9-11_truth_deniers_are_crimina.html.

Moss, Phil. "Leisure Time—The Time To Relax." *Lawrence* (Kansas) *Journal-World,* May 1, 1985.

New Roots. "A New Style of Board Games." *New Roots,* February 27, 1980.

Orbanes, Philip E. *The Game Makers: The Story of Parker Brothers, From Tiddledy Winks to Trivial Pursuit.* Boston, MA: Harvard Business School Press, 2004.

Poser, Stefan. "Leisure Time and Technology." *European History Online,* September 26, 2011, ieg-ego.eu/en/threads/crossroads/technified-environments/stefan-poser-leisure-time-and-technology.

Richards, David. "The Beatles—Illuminati Mind Controllers." HenryMakow.com, August 5, 2012, henrymakow.com/beatles_were_mind_control.html.

*Simpsons-Sears Christmas Wish Book.* Regina, Canada: Simpsons-Sears, 1979.

Tarpey, John P. "Selchow & Righer: Playing Trivial Pursuit To the Limit." *Business-Week,* November 26, 1984.

Walsh, Tim. *Timeless Toys: Classic Toys and the Playmakers Who Created Them.* Kansas City: Andrews McMeel Publishing, 2005.

### 15. Pandemics and Terror: Dissecting Geopolitics on Cardboard

Heymann, David L. "How SARS Was Contained." *New York Times,* March 14, 2013, www.nytimes.com/2013/03/15/opinion/global/how-sars-was-contained.html?_r=0.

Hunt, Katie. "Sars Legacy Still Felt in Hong Kong, 10 Years On." *BBC News,* March 20, 2013, www.bbc.co.uk/news/world-asia-china-21680682.

Shaw, Jonathan. "The Sars Scare." *Harvard Magazine,* March/April 2007, harvard-magazine.com/2007/03/the-sars-scare.html.

Taylor, Chris. "The Chinese Plague." *World Press Review* 50, no. 7 (July 2003), www.worldpress.org/Asia/1148.cfm.

"WHO Chief Promises Transparency on Ebola Failures." *Associated Press,* October 20, 2014, bigstory.ap.org/article/98181b6515da4baa94e6ac53ce880c07/who-chief-promises-transparency-ebola-failures.

World Health Organization. "Severe Acute Respiratory Syndrome (SARS) Multi-Country Outbreak—Update 6." News release, March 21, 2003, www.who.int/csr/don/2003_03_21/en.

### 16. Made in Germany: Catan and the Creation of Modern Board Games

Catan. "World Record! 1040 Gamers at the Catan—Big Game," Catan (blog), October 13, 2015, www.catan.com/news/2015-10-13/world-record-1040-gamers-catan-big-game.

Curry, Andrew. "Monopoly Killer: Perfect German Board Game Redefines Genre." *Wired,* March 23, 2009, archive.wired.com/gaming/gamingreviews/magazine/17-04/mf_settlers?currentPage=all.

Friedhelm Merz Verlag. "Press Information No. 1: International Games Event Spiel '15 With Comic Action From October 8th to the 11th, 2015." News release, July 2015.

Glonnegger, Erwin. *Das Spiele-Buch.* Uehlfeld, Germany: Drei Magier Verlag, 1999.

Green, Lorien. *Going Cardboard.* Film documentary, T-Cat Productions, 2012.

Herz, Jürgen. "Geschichte," Spiel des Jahres, n.d., www.spieldesjahres.de/de/ueber-uns.

Hutton, Ronald. "The Vikings Invented Soap Operas and Pioneered Globalisation—So Why Do We Portray Them as Brutes?" *New Statesman,* February 25, 2014, www.newstatesman.com/ideas/2014/02/vikings-invented-soap-operas-and-pioneered-globalisation-so-why-do-we-depict-them.

Janzen, Olaf. "The Norse in the North Atlantic." Heritage Newfoundland & Labrador, 1997, www.heritage.nf.ca/articles/exploration/norse-north-atlantic.php.

Parlett, David. "Hare and Tortoise." N.d., www.parlettgames.uk/haretort.

Pinker, Steven. *The Better Angels of Our Nature: The Decline of Violence in History and Its Causes.* London: Allen Lane, 2011.

Ravensburger. "Ravensburger Story." Unpublished paper, 2011.

Roeder, Oliver. "Crowdfunding Is Driving a $196 Million Board Game Renaissance." *FiveThirtyEight,* August 18, 2015, fivethirtyeight.com/features/crowdfunding-is-driving-a-196-million-board-game-renaissance.

Scherer-Hoock, Bob. "Evolution of German Games." *Games Journal,* May 2003, www.thegamesjournal.com/articles/GermanHistory2.shtml.

Stubbs, David. *Future Days: Krautrock and the Building of Modern Germany.* London: Faber & Faber, 2014.

"Viking Law and Government: The Thing." *History on the Net,* August 14, 2014, www.historyonthenet.com/vikings/viking-law-and-government.html.

von Richthofen, Esther. *Bringing Culture to the Masses: Control, Compromise and Participation in the GDR*. New York: Berghahn Books, 2009.

Woods, Stewart. *Eurogames: The Design, Culture and Play of Modern European Board Games*. Jefferson, NC: McFarland & Company, 2012.

# ACKNOWLEDGMENTS

While it's my name on the cover, like any book it wouldn't have been possible without the help and support of many other people.

As always the biggest thank you goes to my husband, Jay, for his support, suggestions and—above all—patience.

Thanks too to my ever-fabulous agent Isabel Atherton of Creative Authors, Anne Brewer for signing the book, my editors Emily Angell, Lisa Bonvissuto, Melanie Fried, Jennifer Letwack, and the rest of the team at Thomas Dunne Books. Without them you wouldn't be reading this now.

An extra special thank you goes to Dougal Grimes, door opener extraordinaire. This would have been a lesser book without your help and generosity. Thanks, Keith, for introducing us.

A big thank-you also goes to everybody who took the time to answer my questions and/or help me track down people and information I needed: Nicole Agnello; Sarah Andrews of the Woodrow Wilson House's Vintage Game Night; Joel Billings; Jane Bowles; Gerry Breslin; Jeffrey Breslow; McKell Carter; Charles of Washington Square Park; Rémi Coulom; Alex de Voogt; Lewis Deyong; Jon-Paul Dyson; Tom Felber; Jon Freeman; Jennifer Gebhardt; Mike Gray (thanks for introducing me to Acquire too); Ananda Gupta; Reyn Guyer; Tracy Harden; Heinrich Hüentelmann; Reuben Klamer; Matt Leacock; Jim Lowe; Henry Makow; Jason Matthews; Greg May; Dominique Metzler; Burt Meyer; Steve Meyer; Helen Newstead; Mary O'Neill; Philip E. Orbanes (and the rest of the Winning Moves team); Beatrice Pardo; Chris Pender; Judit Polgár; Ben Rathbone; Volko Ruhnke; Tristan Schwennsen; Chiva Tafazzoli; Benjamin Teuber; Klaus Teuber; Tim Walsh; Bruce Whitehill; and Rhea Zakich.

Cheers too to the archives teams at Brooklyn Museum, the Strong Museum in Rochester, New York, and the West Yorkshire Archive Service.

Finally, thanks to Petra Tank for her sterling German-to-English interpreting efforts, my family for all those years spent playing board games, the Lewes Board Games group, and you, the reader, for buying and reading this book.

# INDEX